Neurobiology of Fear, Anxiety and other Emotions

by

Aage R. Møller, PhD (DMedSci)

Professor of Cognition and Neuroscience
The University of Texas at Dallas,
School of Behavioral and Brain Sciences

AAGE R MØLLER PUBLISHING, DALLAS, TEXAS 2019

iv

Neurobiology of Fear and other Emotions
By Aage R. Møller, Ph.D. (D. Med. Sci.)
The University of Texas at Dallas
School of Behavioral and Brain Sciences,
800 W. Campbell Road
Richardson, TX 75080

Front cover by Zahra Akhavi.

Dr. AAGE R. MØLLER is a Distinguished Lecturer in Behavioral and Brain Sciences and he is the Founders Professor of The University of Texas at Dallas School of Behavioral and Brain Sciences. He has been on the Faculty since 1997.

Dr. Møller has a Doctor of Medicine degree from the Karolinska Institut, (School of Medicine), Stockholm, Sweden. He was on the faculty of the Karolinska Institut for 12 years, on the faculty of the University of Pittsburgh School of Medicine for 19 years, first as Associate Professor of Otolaryngology, and later as Tenured Professor of Neurological Surgery.

He teaches courses on "Biology of Pain", "Human Functional Neuroanatomy", "Sensory Physiology", "Neural Plasticity and Disorders of the Nervous System" and "Intraoperative Neurophysiologic Monitoring, Part I and II" in the Neuroscience Program of The University of Texas at Dallas School of Behavioral and Brain Sciences.

ISBN: 978 1081 392192 © Aage R. Møller, Publishing, 2019

Acknowledgements

During the writing of this book, I had many valuable comments and suggestions from members of the School of Behavioral and Brain Sciences, and I want to mention Dr. Theodore Price, especially. Drafts of this book have served as the text for my teaching a course on Fear, Anxiety, and other Emotions in the Neuroscience Program of the School of Behavioral and Brain Sciences. Many of my students have provided valuable suggestions regarding this book. I want especially mention two of my students, who contributed to the book in several ways. Liliana Ivonne Cabrera Menjivar did most of the illustrations in the book, improving the value of many of the illustrations. Wade C. Deckard, edited the entire text making the book more readable; he also provided many valuable suggestions regarding the book. The front cover is a creation of Zahra Akhavi. I would not have been able to write this book without the support and encouragement of my wife, Zahra Akhavi. The School of Behavioral and Brain Sciences at the University of Texas at Dallas gave much valuable support.

Dallas, August 2019
Aage R. Møller, Ph.D. (D. Med. Sci).

Other recently published books by this author:

Møller, A.R. "A new Epidemic: Harm in Health Care. – How to make rational decisions about Medical and Surgical Treatment" Nova Publishers, New York, 286 pages, 2007.

Møller, A.R. "The Malleable Brain." Nova Publishers, New York, 2009

Møller, A.R. "Intraoperative Neurophysiologic Monitoring, Intraoperative Neurophysiology" 3rd Edition, Springer, New York, 403 pages, 2010, ISBN-13: 978-1441974358

Møller, A.R. "Hearing: Anatomy, physiology, and disorders of the auditory system," 3rd Edition, 415 pages. Plural Publishing, Inc. 2012. ISBN-13: 978-1-59756-427-4, ISBN-10: 1-59756-427-3.

Møller, A.R "Sensory Systems," 2nd Edition. Aage R. Møller Publishing, Dallas 2012. 417 pages, July. 2012. ISBN-13: 978-1478175872, ISBN-10: 1478175877.

Møller, A.R. "Pain: Its Anatomy, Physiology, and Treatment," Aage R. Møller Publishing, 2018, 364 pages; 2nd Ed. 2014, 403 pages, ISBN-10: 1466395109 | ISBN-13: 978-1466395107

Møller, A.R. Neuroplasticity and its Dark Sides: Disorders of the Nervous System, 2nd Edition. Aage R. Møller Publishing, Dallas, 2018, 244 pages, 2018. ISBN-13: 978-1981464906 ISBN-10: 1981464905

General Introduction

Fear plays a significant role in almost every person's life. Fear is the belief that something dangerous or unfortunate may happen in the future. Fear controls people's reaction to many aspects of life. Fear plays a huge part in the body's defense or protection mechanisms with many similarities to other perceptions that serve to protect the body from harm. Fear may help someone avoid making harmful choices, but it may also make people make decisions that are not beneficial. Fear, in itself, is regarded to be an unpleasant perception with few exceptions such as the pleasure or joy that people may experience from watching horror movies.

Fear is a part of an extensive defense system that warns about a danger to the body. Fear can prepare the body for reactions that can protect an animal or a person from threats or danger by mobilizing the resources of the body for fight or flight. Fear provides a warning that something is dangerous. Fear may make a person avoid or take extra precautions from injuries situations, illness, or losing his/her livelihood from loss of money and other assets. Fear is one of a few innate emotions.

Anxiety is a general term for several disorders that cause nervousness, fear, apprehension, and worrying. Anxiety is different from fear. Fear is related to specific perceived dangers, while anxiety is an emotion that occurs as a result of the perception of uncontrollable threats or circumstances. Anxiety is related to specific problems such as an exam, important decisions, or it can be generalized. People with generalized anxiety disorders (GAD) have an excessive amount of anxiety most of the time for long periods. These disorders affect how a person feels and behave, which can cause physical symptoms. An anxious person is concerned about many more things than a person who is not anxious.

Worry is another often-used general term that means to be concerned or nervous. A person may worry about things that are very unlikely to happen or things on which a person has no influence. People may worry about the result of an upcoming governmental election. Voting may relieve some worry, but anxiety promotes and exacerbates apprehension.

Fear that is directed to perceived dangers that are indeed harmful is beneficial if the person in question takes action that can reduce the risk of occurrence of the dangerous event or situation, or, if it occurs, reducing the harm. For example, one would think that the risk of getting a severe disease would cause enough fear to get people to seek protection in the form of vaccination or take action to avoid exposure to the virus that causes the disease.

However, statistics show otherwise. For example, only a small fraction of people in the USA take the time to get a vaccination against diseases such as influenza, and for seniors, pneumonia. Too little fear makes people embark on reckless activities such as driving recklessly, wild speculations on the stock market, or sex without taking adequate precautions.

Fear can be harmful in itself, such as when it prevents a person from achieving something that could have been beneficial to the person. Fear can be dangerous when harmless things or events unlikely to occur scare people. Many rules, such as social norms or religious protocols, induce various degrees of hardship on people. Such rules are only followed, to a great extent, because people are afraid of the consequences of not following the rules.

This means that fear keeps people from breaking the rules, and the rules are not followed because they offer a person a benefit. Fear can paralyze an entire nation in situations such as in war. Famous political leaders have made statements about fear. "Irrational fear is a major impediment to success and productivity. When Franklin D. Roosevelt acknowledged, in 1933 "the only thing we have to fear is fear itself," he was commenting on the economic future of the United States, but unreasonable, over-generalized fear can have dramatic effects on all aspects of one's life" {Mahan, 2012 #6007}. Winston Churchill said: "Fear is a reaction. Courage is a decision."

Fear, or lack of fear, controls many people's actions or inactions, and it controls the action of animals as well. Fear can induce actions that can protect the body; fear can initiate change in many vital body functions that are involved in the initiation of the fight or flight mode of the body that acts as a protection in emergencies. Fear of diseases may promote adaptation to a healthy lifestyle to help reduce the risk of getting sick and encourage a person to take other precautions such as vaccination and more.

The fear of having a traffic accident can make a person drive more defensibly, the fear of becoming pregnant can prevent people from having pre-marital sex, and the fear of addiction to recreational drugs can make a person ovoid trying addictive drugs.

Fear often causes changes in essential body function such as activation of the autonomic nervous system, increasing the sympathetic activation, and decreasing the parasympathetic activation. Some forms of fear may cause a person's heart to beat faster, whereas other forms of fear do not do that or do it to a much less degree.

Many authors have discussed how fear may impair a person's quality of life. Here is an example:

"I must not fear. Fear is the mind-killer. Fear is the little-death that brings total obliteration. I will face my fear. I will permit it to pass over me and through me. And when it has gone past, I will turn the inner eye to see its path. Where the fear has gone, there will be nothing. Only I will remain". Frank Herbert, Dune (Herbert, 1964.)

Fear is based on predictions about what could happen in the future. Predictions of what may happen in the future are often incorrect or fail to materialize at all. Fear is often misdirected and causes people to worry unnecessarily. Fear mostly depends on one's beliefs and often ignoring the facts. This makes many forms of fear misdirected and people, in general, are afraid of the wrong things. Unsupported and unfounded fear may cause harm because, in itself, this kind of fear is harmful, and unfounded fear may make a person make choices that are, ultimately, not beneficial. Unfounded fear may be the most basic form of fear, but it has attracted little attention.

Fear can be caused by an event that occurs suddenly or, more commonly, fear is often caused by the expectation that some specific event poses a danger of one kind or another and will occur in the future. By far, most of these events will never occur, and the unfavorable consequences of many of those that may occur can be prevented or reduced by appropriate preclusions. Anxiety has similarities with fear, but often, it concerns unspecific causes.

The beneficial effect of fear depends on the actions a person takes in response to a perception of fear. If a person fears being sick yet takes no action, then there is no benefit from the fear. There is only its adverse effect of reducing the quality of life that remains. A person can ignore to react to the fear he/she experiences, react in proportion to the amount of danger he/she may experience, or overreact, act much stronger than justified by the danger in question. Appropriate action in response to fear should take into account the likelihood that the undesirable event will occur and the consequences that are experienced when it occurs.

Fear and anxiety are emotions that concern the functions of the brain. Understanding fear and anxiety, therefore, depends on understanding the function of some of the complex systems in the brain, belonging to the limbic system, where the amygdala plays an important role. It is only relatively recently that studies of neurobiology have illustrated the general functions of these parts of the brain.

For example, it is hard to understand many aspects of theoretical physics, but we can understand that the big bang theory is about the creation of the universe and ultimately the creating of the earth and all living material including us.

The different features of psychology and neurobiology of fear and anxiety are examples of complex problems that are difficult to comprehend. To some extent, because of limited knowledge about the function of complex neural systems. What risks a person should be concerned about and to what extent a person should take actions to reduce the risks of a particular adverse event are more mundane questions that are manageable.

Many people fear matters that do not pose any noticeable likelihood of causing any harm, but there are also risks that have a high probability of causing harm. Some potentially harmful happenings do not induce a sufficient amount of fear resulting in a person taking appropriate actions to avoid or reduce the risks.

There may be different reasons why a person ignores to take appropriate actions; some may be related to the fact that only the likelihood of the occurrence of many harmful events can be determined.

Many people are unable to imagine the risk of dying in an automobile accident is far higher than it is flying on a commercial airline. Many people cannot even imagine that they can catch influenza and therefore do not take a beneficial action that can reduce the risk.

This book was written to promote a balanced understanding of fear and anxiety. The book shows that many lives could have been spared and many people could have been saved from suffering a reduced quality of life by better management of the emotion of fear.

The first chapter discusses the basics of fear. It discusses the various definitions of fear and how to approach an understanding of fear. The second chapter covers some basic aspects of fear and anxiety. What is fear, what can cause fear, and what can fear do to a person? It mentions different forms of threats that can evoke fear and how anxiety can modify a person's perception of fear. How fear and anxiety affect essential body functions and other human functions, such as cognitive abilities are also discussed.

The first chapter also discusses the causes of fear, the nature of the different forms of fear, and how the body (the brain) handles fear and anxiety. It explains what is real, and what is imaginary exaggerations and untrue causes of fear and it discusses the benefit and harm from fear and anxiety and how to manage some common risks of disease, trauma and other forms of harm.

It is well known that perceived fear is poorly related to real risks. This is the case even when a person knows the real risk. This means that fear consistently overrides and manipulates facts. There is evidence that this occurs in specific parts of the brain, mainly involving the amygdala and other structures that belong to the old brain.

The second chapter covers basic aspects of emotions. It also discusses the components of a good quality of life and the prerequisite for success in life. It discusses the variations of the theory of James Lange and the role of various forms of an appraisal. The role of a person's core values is discussed in connection with a person's quality of life.

The third chapter discusses the neurobiology of emotions. The historical development of an understanding of the function of the "emotional brain" is discussed. Chapter 3 also outlines how sensory input controls fear and how the amygdala nuclei play the primary role in creating the action to be taken from fear signals. The role of memories and imaginations in activating emotions such as fear and anxiety is also discussed in Chapter three.

What specifically happens in the brain when a person experiences fear or anxiety is discussed in the fourth chapter of the book. Many brain systems that are involved in fear and anxiety may be activated, primarily including the three nuclei of the amygdala, but also the insular lobe and the anterior cingulate cortex are heavily involved.

The fifth chapter discusses explicitly the neurobiology of threats. What to be afraid of and what not be afraid of is also discussed in this chapter. Fear is a part of the body's alarm system that can warn about the dangers of various kinds that may occur sometime in the future. Fear can be a warning signal about various kind dangers and it can initiate actions that can prevent (or rather reduce the risk of) damage to the body. Signs of danger may also come from inside the body in the form of signs of diseases or signs of a need for fluid (thirst) or nutrition (hunger).

The sixth chapter discusses how to minimize the risk of illnesses and other forms of bodily harm, and it discusses what to be afraid of and what not be afraid of. Fear is a part of the body's alarm system that can warn about the dangers of various kinds that may occur sometime in the future. Chapter 6 discusses the benefits of preventive measures such as vaccination and the reasons that these benefits are not fully utilized. Some of the reasons for misuse of medications such as opioids are also discussed in this chapter.

Table of Contents

Chapter 1 Fear and Anxiety

Introduction

Fear and anxiety are emotions that are caused by a person's belief that some unpleasant, harmful, or dangerous event may occur in the future and that it may affect the person in adverse ways. Such dangers may come from outside of the body in the form of threats of various kinds or expressed internally.

Fear can be described in many different ways such as concern, apprehension, uneasiness, unease, fearfulness, disquiet, disquietude, inquietude, perturbation, agitation, angst, misgiving, nervousness, nerves, worry, tension, or tenseness. A person can be afraid of real threats such as death, or from imaginary threats such as that of God or Hell. Fear can be exaggerated or directed towards harmless matters, or fear can be elicited by the belief that a very unlikely happening will occur. Fear and anxiety can be caused by factors that are not signaling danger at all. This is what is called irrational fear.

Fear is one of the six innate emotions: anger, disgust, fear, happiness, sadness, and surprise. These are considered the *primary or basic* emotions and also referred to as *archetypical* emotions. Like other emotions, fear can affect a person's quality of life, and it often affects many essential bodily functions such as autonomic functions. These autonomic functions can affect a person's blood pressure, heart rate, perspiration, and muscle tremor. Many bodily functions such as the immune system, the vascular system, and the endocrine system may be affected by fear. Fear affects a person's mood and a person's quality of life in general. Strong fear may shift the body into a fight or flight mode.

This chapter discusses the different forms of fear and anxiety. The difference between rational fear, and irrational fear and anxiety are also topics of this chapter as are the different causes of fear and anxiety and what fear and anxiety can cause. Also discussed are some practical aspects of fear and anxiety such as the choice of what to fear.

Fear is a part of the body's alarm system that can warn about the dangers of various kinds that may occur sometime in the future. Fear can warn about dangers of the various kinds and initiate actions that can prevent damage to the body. Signs of danger may also come from inside the body in the form of signs of diseases or signs of a need for fluid (thirst) or nutrition (hunger).

Too much fear may perturb a person unnecessarily and reduce the quality of life for a person. This will consume a person's time and wear unnecessarily on the body. On the other hand, too little fear may make people embark on reckless activities such as driving his/her automobile recklessly, making wild speculations on the stock market, or having sex with people he/she does not know. Fear does not reflect exactly what dangers may occur and when the dangers will occur, but fear only concerns the likelihood of the occurrence of untoward events.

Most of what a person can be expected to encounter in life cannot be determined accurately, but only the likelihood of the outcome can be predicted or guessed. The stochastic "chance effect" is from the Greek word στόχος (stokhos) meaning "aim, guess". Most of the events that involve accidents and diseases are random events about which it is only possible to guess (predict) that a specific event will happen in the future and when it will happen.

Only a few things can be predicted accurately, and only the likelihood of occurrence of most events can be determined. In fact, the only events that can be predicted accurately are the movements of celestial bodies including the sun and the moon. While a person's death is inevitable, only the likelihood of when it will occur can be determined. Fears from potentially dangerous issues that are very unlikely to materialize must be evaluated regarding the probability for them to occur and the potential consequences. If the consequences are minimal, most people may find it an unnecessary fear, but if the consequences are catastrophic, most people will take precautions even for unlikely events.

Ideally, the amount of fear a person experiences should be directly related to the risk of harm, the likelihood it will occur, and the severity of the consequences of the harm that is caused if the fear is expressed. In general, however, a person's fear is often determined by the person's belief of what is dangerous and what is not dangerous and the expected likelihood of occurrence, but beliefs, as we already know, are rarely related to facts and knowledge. In fact, the belief of expected harm is often (very) different from the real risk of harm.

In another situation, fear is caused by the combination of several factors that must exist together to elicit a perception of fear. This is similar to many other common things such as diseases and the causes of a person's death, which is often a combination of many factors.

Many people fear to be old or to be sick or dying. People with children fear something terrible might happen to their children. People fear losing money or not having enough money to pay bills or to fulfill their wishes. Married people can fear that a spouse becomes ill and dies, or that he or she is having an affair. Some married people are afraid of their spouse because of the threat of physical violence or mental harassment of various kinds.

The role of belief in creating fear

Fear is caused by the belief that a dangerous happening may occur and affect a person by causing harm or unpleasantness in one way or another. Often the belief of a risk of harm is not based on any concrete facts, but a person's opinion is often stronger than most facts. What a person believes cannot be proven, and it is therefore not disputable. Many people's beliefs about different dangers are much stronger than any evidence.

This is what makes people have exaggerated fear and be afraid of matters that do not pose any danger at all or which poses minimal risks. How are people's beliefs created? Ideally, one would think that what a person believes would be the sum of many experiences and knowledge acquired from many sources. In reality, fear is often nothing but related to real danger. Unfortunately, a person's belief often overrides facts that are generally and widely available.

Fear is often based on belief and not on facts. A person's fear is most often directed to a danger that never occurs because their prediction about what is dangerous is mostly wrong. This means that most people's worries are in vain. So why worry about a specific matter when experience clearly shows that it is almost impossible to predict what will happen in the future.

Ideally, fear should be related to knowledge about the likelihood that something that is harmful will occur.

In *reality*, people often fear matters that cannot be proven or disproven to pose any harm. Both real and imaginary threats can cause fear.

In our modern society, one would expect that decisions are made from known facts, but important decisions are often based on beliefs, and those beliefs can easily override the facts. In modern times, there are many situations where fear promotes effective actions that could reduce the risk of diseases, accidents, and other adverse events. Unfortunately, it seems to be a more familiar experience that fear produces extreme reactions that are not related to the amount of risk or the consequences if the risk is manifested.

Rational and irrational fears

The basis for the feeling of fear has evolved to protect biological organisms from external dangers in hostile environments, and these dangers may include intrusion from microbial pathogens, physical and chemical threats, and from internal threats such as poisons that have reached the body and from illnesses of various kinds.

Perception of fear

Decisions regarding how to react to dangers are often based on a person's **perception** of the risk of harm and not on the **actual** likelihood that something harmful would occur. Unfortunately, an incorrect perception of risks often controls or direct a person's choice.

Different people describe fear differently. This is akin to the Indian fable about the six blind men who describe an elephant. The men could not accurately describe the elephant because they did not have all of the information. Each one made assumptions based on the incomplete information they had. There is no doubt that a person's education, occupation, and professional background affect the person's view of fear. A psychologist, a psychiatrist, a neuroscientist, a social worker, an economist, and a politician all will have distinct views on fear, but their views may differ. A neuroscientist is more likely to emphasize the role of fear as one of the several mechanisms that protect the body, and many neuroscientists may focus on the different kinds of reactions their severity and the form the different fearful situations have. A psychologist may regard fear to be an emotion or experience that can be induced by either physical threats or by imagined threats. Psychologists often regard fear and love as the two strongest motivating forces in life.

Many people enjoy the fear that they can experience from viewing horror movies or from riding a roller coaster. Watching a horror movie may produce a feeling of danger, but there is no risk associated with that fear. This is because the movie is not real and entirely harmless. The fear a person experiences from a threatening person with a weapon who wants to rob him or her is very different from that experienced when watching a horror movie.

Fear can be purposeful or not purposeful

Fear is purposeful if it identifies a specific risk. Fear is not purposeful if directed to the wrong matter or if no specific target can be identified. If fear is directed to matters that pose no risk, the fear may be purposeful but not beneficial. An example is a person being afraid of harmless objects or harmless events.

Fear can be purposeful if it is able to identify the source of the fear correctly, but it is not beneficial regarding matters that imply no risk of harm, or if the fear is directed to matters that cannot be changed. There is "good" fear and "bad fear." Fear that helps identify and remedy harm that can counteracted is "good" fear. Fear that is harmful but the cause of which cannot be found, or the harm cannot be remedied is "bad" fear. This is similar to other primary signals from the body such as pain. Also, other functions such as the ability to learn and acquire new skills (through activation of neural plasticity) have both beneficial sides and harmful (dark) sides [1].

Phantom risks

While some of the risks most people are concerned about is either phantom risks or risks that a person cannot change, there are a sufficient number of real risks that a person can affect to satisfy the desire of most people to have something real to worry about. One should leave room for taking adequate preventive measures for all the risks where a person can reduce the risks. For a start, take a vaccination for influenza, be sure to always wear seat belts in automobiles.

A person's fear may be induced by information stemming from many of the spectacular events shown on television, and many are so rare that the risk it would occur to a person is minuscule.

Rational fears

Rational fears include fears of real matters such as harm to a person's body or damaging effects on a person's everyday life or loss of property. The risk that a person gets a severe disease may also elicit a rational fear. A person may have this kind of fear when having pain because the pain may be a sign of a disease or an unseen threat. People may fear walking on the street when it's dark because of the risk of robbery, or a woman may be fearful because of the risk of being raped. People may fear becoming old because that may imply that a person cannot take care of himself/herself.

Being afraid of driving a personal automobile in heavy traffic or flying on commercial airlines are examples of rational fear because it involves a real danger to a person, and the risk can be avoided or reduced if actions are taken by a person. However, there is an important difference in the risk. Traveling almost always involves risk but choosing the way of travel can reduce the risks. Traveling by commercial airlines possesses a much smaller risk than traveling by a personal automobile.

The risk of having an accident while driving a personal automobile is far greater than traveling on an (American) commercial airliner. In fact, it may be claimed that traveling on commercial airlines may involve less risk than doing everyday activities such as walking on a street or in one's own home. Not everybody can take advantages of the possibilities to reduce these risks effectively.

Worry about getting a severe disease is a rational fear if there are known precautions that can be taken to reduce the risk of getting the disease. The fear of falling in the home, workplace, or outside affects many people by causing death and trauma. Known effective precautions can reduce the risks of falls. Vaccinations are now available for many diseases. Many people are scared of being a victim of crime.

There are many examples of rational fear. Here two examples of common rational fear that affect many people, fear of illness and fear of falling.

Fear of illness

Most people are getting ill at some time, and many people are afraid of getting specific diseases. These worries are often not based on facts such as hereditary aspects, and therefore most concerns rarely materialize, and instead, a person will most likely have different diseases than those he/she is worried about. Worries about getting sick are only of value if it is possible to take precautions that will reduce the risk of getting sick. There are now many ways of doing that. Vaccinations are very useful in reducing the risk of many different infectious diseases, and it saves many lives and spares many people from the suffering from illnesses.

Fear of falling

Many people are afraid of falling. Especially old people or people who have had a fall that caused injuries are afraid of falling. Sometimes the fear can be so intense that it interferes with everyday life and it may prevent a person from leaving the house because they are afraid of falling again.

Fear of falling can increase the fall risk, especially in older adults [2]. It is widely reported that fear of falling (FOF) is prevalent in older people. There are many causes, many of which have no remedy such as age. Some problems, such as balance related problems may be resolved. Poor shoes is a problem that is easy to solve. People being aware of the problem is also important and can reduce the risk of falling. Falls in bathrooms is a common cause of hip fractures and head injuries; both have severe consequences. Wet floors such as in the shower is a common cause, and the risk of that can be reduced by placing a towel on the floor of the shower. Putting a carpet on the floor in the bathroom decreases the risk of falling. Falls often occur when a person is in a hurry. Therefore, taking time to walk carefully reduces the risk of falls and thereby the risk of injury and severe deterioration of a person's health. Probably many people know all that, but for various reasons do not fix these problems before someone falls.

Fear of people who are different (xenophobia)

Many people discriminate or dislike people who are different from themselves whether it be different skin color, intelligence, sexual orientation, place of birth, or something else. This is a human made fear because people who are different are not a threat to other people because of these differences. People who are different may naturally be a treat like anybody else.

Xenophobia is a term that describes a widespread and diverse phenomenon (dictionary: fear and hatred of strangers or foreigners or of anything that is strange or foreign).

Such discrimination causes fear and anxiety of people being discriminated. People who are not religious probably believe that Darwin was right and that all living creatures were created through evolution. This means that the same evolution that created heterosexuals also created homosexual people. Thus, there is no reason to fear people who are different in one way or another. People who are religious would believe that creation was done by God and that it would follow the same rules for the same origin of all people and there would therefore be no justification for discrimination based on differences.

Different religions assume that their Gods is without any error and does everything right. People, in general, also believe that the God in which they believe has created everything including all humans. This means that it is to be assumed that the same God has created all people independent of their sexual preference, the color of the skin, their intelligence, or wealth. It is, therefore, strange that some religious people have an intense dislike of gay people. Similar reactions seem to be present against people who have changed their sexual identity. That is naturally a source of fear.

Fear also exists in animals. In a rat colony, a rat that has been hurt or operated upon is killed by the other rats. In addition, there are many examples of people who fear the abnormal. The fear expressed as hate is seen, for example, against people who have "abnormal" sexual preferences such as gay men and lesbians. The hostility against people who have a different sexual preference than what is regarded to be normal largely originates from religious organizations, but the fear of individuals who are different may have a biological reason (similar to xenophobia).

Some risks are very small

The amount of fear that specific threat or danger creates varies widely among different people, and it varies throughout the day or under different circumstances for any given person. The degree of fear that a person may experience is often either much stronger than justified by the risk or much less and only rarely is a person's fear proportional to the severity of the danger and the likelihood that the danger will occur.

Many risks are so small that the risk of being affected for a person is extremely low, similar to winning in the lottery or lower. The Ebola virus was a scare for people in the USA when one person with the virus died. Shortly after, polls showed approximately 80% of the people who live in the USA feared to catch the disease (compared with the average deaths from the universal influenza of 20,000 in the USA per year).

Irrational fear

Irrational fear is the fear of matters that cannot cause any harm or inconvenience or signal damage of any kind to a person. Irrational fear is the fear of harmless matters or things that only exist in a person's imagination. Irrational fears also include fears that are exaggerated beyond what is realistic. Irrational fears can be all kinds of fear of harmless matters or happenings that cannot cause any harm or threat of harm.

Fear of matters that a person cannot affect is also irrational because it lacks any forms of importance. That kind of fear is different from rational fear, but it activates the same body systems as a rational fear does. Being afraid of totally harmless things, such as a salamander, is a form of irrational fear. Being afraid of spiders is also regarded as a classical source of irrational fear.

Many people fear matters that cannot possibly cause harm, and which are unlikely to affect a person in any negative way. Here are some examples: Darkness, public speaking, heights, and more that cannot possibly inflict any harm. The fear of God and Hell are other examples of fear that is not based on objective observations.

Public speaking is a familiar source of fear, but the risk of harm seems to be minimal and the consequence may be a feeling of embarrassment at the most. Some people experience fear when in an uncertain environment. Yet, if there are no specific dangers in the environment, then this would be an irrational fear. The elephant that was scared by a mouse is perhaps the best illustration of what causes many people to be scarred for no rational reason.

Fear is closely related to religions and the fear of God is universal. In fact, the English translation of the word used to describe a religious person in many languages is that he or she "fears God" (God-fearing). Fear of God seems equally irrational because nobody has ever been able to present evidence that God could do anybody any harm. Despite these irrational causes of fear of God and other religious beliefs, religions can get people to do unbelievable things causing sufferings and inconvenience. Likewise, many people fear they may have to go to a place known as Hell after death. There is no scientific evidence that Hell exists at all and that a person should suffer the afterlife in an unpleasant place lacks rational support. Hell-like conditions do indeed exist on Earth, as do heaven-like conditions.

This means that often people's worries are in vain. Consequently, it is a waste of effort to worry about most of the things that people commonly worry about. So why worry about a specific fear inducting matter when experience clearly shows that predictions of what will happen in the future are inaccurate at best? Many people worry about matters they cannot control, and that is equally unrealistic. Taking appropriate precaution is another matter that can be beneficial.

Many people are afraid of the wrong things

For example, many people are more afraid of a terrorist attack than of catching influenza. Another everyday example of poor decision making is related to how people travel. Many people select to travel in a personal automobile instead of flying on a commercial airline because they truly believe that traveling by automobile is safer than flying. This belief is untrue. One of the reasons for those misconceptions is that people in general have an overstated estimate of their skill as a driver of their own automobile while at the same time they are doubtful about the skill of bus drivers and airline pilots.

The saying, "It ain't what you don't know that gets you into trouble. It's what you know for sure that just ain't so", by Mark Twain applies to matters related to fear and anxiety.

Many people are doubtful about automated system although usually associated with much less risks than human operators.

In general, many people worry about circumstances that have very little or no possibility to do harm to a person while disregarding obvious causes of harm. People also often worry about things they have no control over such as the country's government while disregarding things that are obviously harmful and within a person's control such as vaccination or lifestyle.

Misdirected fear (worries)

There are many forms of misdirected fear or worries. The word, worries, describes a kind of misdirected fear that may be harmful to the body and the mind. These worries cause stress and can place the body in an alarm mode. Misdirected fears or worries decrease the quality of life.

Here is a list of common misdirected worries:

1. Worry about something that cannot be proven or disproven such as the existence of Hell (believed by at least one billion people).

2. Worry about terrorism in the USA, while the risk of being harmed by an intoxicated driver is a much higher risk.

3. Fear of getting cancer from environmental causes when the risk of getting cancer from a person's lifestyle is much higher.

4. Worry about flying on commercial airlines while the risk to die in an automobile accident is at least 50 times greater for the same distance traveled; the risk to become seriously injured in an automobile accident is several hundred times greater than from flying overall.

5. People ignore to take precautions or get preventions for diseases (such as vaccinations) that often are by far more effective than getting treatment after contracting the disease and has fewer side effects.

6. Worry about contracting rare diseases such as the one caused by the Ebola virus, even though the risk is minimal to a general person and the risk of catching a common disease such as the influenza is much more significant.

Only the likelihood of the occurrence of adverse events can be determined

Many common daily events can result in adverse effects, but when and if such events will occur cannot be determined, only the likelihood can be determined. A person's choice of what to avoid must, therefore, be based on the likelihood that some adverse effect may occur. However, most people have difficulties in dealing with uncertainties. This is one of the reasons that people often fear the wrong things, and such misdirected fear causes many people to worry unnecessarily and often causes a person to make choices that are not beneficial. A person's choice of what is regarded as dangerous is often based on the person's beliefs. Additionally, people's beliefs are often not based on facts or knowledge about the likelihood that a certain adverse event will occur.

What causes fear and anxiety

External and internal sources of fear perception

Fear can be elicited by sources that are external and internal to the body which are mediated through sensory systems, or from the brain from memory or imagination. The experience of fear can be elicited by a sudden event either through vision or hearing or touch. It can be something we see such as a snake, lightning resulting from thunder, or fire in a house.

Signals from external (environmental) sources that are conveyed to the brain by the sensory systems can cause acute fear and activation of the autonomic nervous system. Signals elicited by somatic and visceral receptors are signals from internal organs such as thirst, hunger.

Injuries to the body and symptoms of a disease can elicit a fear response if the symptoms are interpreted as a sign of a disease. An illness related response mediated by the vagus nerve can likewise cause acute fear for the same reasons. Thoughts and memories can elicit acute fear and chronic fear. Illogically, the unknown is a common cause of fear. Imaginary and unknown phenomena such as Hell are common causes of fear. Harmless matters such as salamanders and various kinds of phobias are causes of fear.

Fear can also be a persistent experience that may be elicited by prior events or through information that infiltrates a person's awareness. Fear may be triggered by what a person sees, hears, or smells, but fear can also present itself as a memory of what is regarded as dangerous events. However, fear is most often triggered by what a person believes is dangerous.

That form of experience may decrease rather quickly after the event that caused it, or it may have a lasting effect as an experience. Fear can be caused by the signs or sudden events that indicate imminent danger. A person may experience fear from the belief of threats of various kinds or by the imagination or memory of prior events. Fear is often caused by a perceived threat.

Fear can also be a part of anxiety disorders with symptoms such as apprehension, general uneasiness, and much more. The imagination of what can happen in the future in terms of frightening events is another cause of fear. Fear can be induced by actual threats or by imagined threats. The severity of the fear a person experience depends on how the threat is perceived. The degree of fear a person perceives in a given situation is often exaggerated or diminished relative to the real severity of the threat.

People can be afraid of flying because they imagine the effect of a crash of an airplane; other people may be afraid of driving because they can image the effect of an automobile accident. Fear is often specifically visualized such as the appearance of a robber or an animal such as a bear. People get afraid of seeing blood on another person, but some people are not afraid of matters that have a high likelihood to occur and where the consequences may be fatal.

The role of belief

Fear may be caused by a person's belief that something unfortunate or dangerous may happen in the future. As well as external dangers, signals from inside the body such as pain and other signs are interpreted and indicate to a person that he/she has a disease and may cause grave fear.

Belief plays a significant role in the perception of what to fear and to what degree. Ideally, beliefs should be based on information, but it is often based on a person's *imagination*. A person's imagination about what possess a danger is often poorly related to reality, such as the history of what has caused a multitude of dangers.

Belief is firm and resistive, and it is often only loosely related to reality or sometimes not at all. Many people's belief regarding fear is often related to incorrect information, and it may be created and spread to other people because of ignorance. Additionally, incorrect information may also be spread to mislead people in one way or another. It is very much possible to get people to believe in untrue matters. If a lie is repeated enough times, many people believe it.

Role of genetics and epigenetics

Like so many other personality traits, the ease with which a person becomes fearful is something a person is born with. This means that the tendency that events or experiences will cause fear in a particular person is, to some extent, genetic. There is nothing one can do to change that; we cannot choose our parents and our grandparents. Fear also depends on epigenetics. Epigenetics is the study of how external or environmental factors switch genes on and off. Epigenetics affect how cells read genes instead of being caused by changes in the DNA. Also, epigenetics refers to functionally relevant changes to the genome that do not involve a change in the sequence if the nucleotide of the DNA. This means that factors from outside the body may influence how a person reacts to fearful situations.

Conscious and unconscious fear

Much more is known about conscious fear that that of unconscious fear, but a few studies have shown indications that fear can be elicited even in response to stimuli that do not evoke a conscious awareness. This means that a stimulus can reach the amygdala even though they do not produce awareness [3, 4] [5].

Others have claimed that such unconscious fear processing depends on a particular subcortical route of input to the amygdala that typically bypasses the cerebral cortex [6-8] [5]. The results of some psychological experiments indicate that there is a possibility of unawareness of the feeling of fear itself (non-conscious emotions), rather than just of the eliciting stimuli [9] [5].

There is now increasing evidence that emotional stimuli that do not produce awareness may induce distinct neurophysiological changes which influence behavior towards the consciously perceived world. Understanding the neural bases of the non-conscious perception of emotional signals can help clarify the phylogenetic continuity of emotion systems across species and the integration of cortical and subcortical activity in the human brain.

Many emotional stimuli are processed without producing awareness. Recent evidence indicates that subcortical structures have a substantial role in this process. These structures are part of a phylogenetically ancient pathway that has specific functional properties and that interacts with cortical processes.

Fear from imagined threats

Fear from imagined threats varies widely among different people. The importance of a person ascribes to a threat also contributes to the severity of how the person perceives it. Fear also depends on a person's imagination of a threat. The severity of the real risk of harm is related to the statistical likelihood that the harm will occur and the severity of the harm such as that to a person's health or life, to social and professional status, power, wealth, and security. However, the amount of fear that a specific risk causes instead depends on the perceived danger which varies widely among different people. That fear within a given person varies during the day and is affected by the circumstances.

Pain is a frequent cause of fear, but only if it is interpreted to be a sign of danger such as a disease or trauma. If a person who wakes you up in the night with pain in the stomach it is likely to cause fear because many people will interpret it as a sign of disease. If the person then finds out that he/she does not have any severe disease the pain may not change, but it does not cause fear anymore. This means that pain only sometimes causes fear.

Causes of fear that are unlikely to materialize

Many people are afraid of terrorist attacks. However, the risk that a person in the USA should die or otherwise be harmed be harmed due to a terrorist related event is extraordinarily small. In 2014, there were 18 deaths in the USA that was terrorist-related, thus giving an averaged likelihood of approximately one in 20 million. This is far less than the risk of being hit by lightning. When a terrorist event occurs somewhere in the World, the news is spread all over the World causing many people to become frightened.

Many causes of fear are unlikely to materialize. For example, people may fear dying in a traffic accident, but will instead die from a fall in the bathroom. Fears that have no or very little likelihood to materialize can still harm some people. People who have that excessive level of worry or anxiety may appear to be suffering in general. The same is the case for people who have excessive degrees of pessimism.

Generally, predictions of the future are unlikely to become true, but this is more pronounced for pessimistic predictions than for optimistic projections. There is, therefore, no reason for spending time and energy on worries about a matter that may sometimes happen in the future.

Human-made fear

Imagination plays an important role in creating fear and anxiety. This is based on the ability of the mind to create and form mental images or concepts that are not actually present and activate certain sensory systems. Imagination such as about what poses a danger is often only weakly related to reality, such as the history of what has caused dangers.

Fear can be used to dominate a person or a group of persons

Fear is probably the most commonly used method to control other people. Fear is used by a person to dominate another person such as when done in a relationship between two people such as in marriages. A threat that creates fear is used by many people to obtain what they want in many connections; from parents' attempts to control their children to employers controlling their employees to politicians controlling large groups of people. Fear is used in general to gain influence over other people. Namely, fear is used for gaining economic or political power. In politics, fear is often used fear to gain people's allegiance.

Fear can be utilized as a weapon to terrorize people like nothing else. Fear is the most efficient of all great manipulators. People can be moved to do anything, no matter how irrational, unscientific, or unwise it may be. Inducing fear is used for personal again and to manipulate other people. The human-made laws of different religions and nations have induced fear in individuals and entire populations with enormous consequences.

Fear of pain and the threat of death is used in torture for obtaining cooperation from a person such as for getting information that the person does not want to give. Waterboarding is a prime example of inducing fear as a means to control a person or get the captive to do what that person does not want to do such as giving information to an interrogator.

Bosses may use fear to dominate a few or many employees; sometimes teachers and professors use fear to dominate students. Parents induce fear in their children to get them to do things they do not want to do, or fear is used to get people to abstain from doing things a person does not want them to do or from doing things an unacceptable way. A person may be afraid of his/her boss because the boss is the person who pays the employee.

A child may be afraid of a parent or their teachers, and students in universities may be afraid of their professors. A politician may tweak the truth to scare people to vote for him/her in an upcoming election by pointing to something that may happen or may not happen if he/she does not become elected.

Governments use fear to have citizens behave in a certain way. Present day political rulers have used methods, such as the fear of terrorism, to gain control of many people.

Fear created in the name of God is perhaps the most powerful means to create fear. Religions has used reward and fear of various kinds to dominate people. The rewards include the expectation of getting an afterlife of pleasure.

Threats such as going to an unpleasant place, Hell, after death are a typical scare that is based on something that cannot be proven. Fear of going to Hell in the afterlife is used by many religions to make people follow specific rules created in the name of God by humans despite there being no evidence that supports these kinds of fear. One would expect religious regulations to be made for the good of people, but it is not always obvious who benefit from these rules. Some of the rules involve a burden for many people, and the fear of violating rules and laws can have substantial effects on many people's daily lives.

Theocratic government rulers make people follow religious rules that are created by people in the name of God. Such rules are more effective than laws created by governments. Kings in the old kingdoms used fear to control their constituents and for personal gain. They got whatever they wanted in pursuit of personal advantages and wealth by creating fear of jail and other forms of political oppression.

Fear mongering is a form of using fear as a weapon (fear mongering: "the action of deliberately arousing public fear or alarm about a particular issue."). It is challenging to reverse the effect of fear mongering because people are uncomfortable with changing their mind.

Some societies impose many different forms of restrictions on their people. Some societies enforce what they want people to do or not to do using the fear of punishment of some kind. Many people live under restrictions about what to think, what to write, and what to say publicly. The ways used to enforce such restrictions are mainly fear of getting punished or fear of not getting a certain reward.

Many years ago, in the USA, the fear of communism affected the government and thus the entire country. It made people fear being regarded as a communist; people feared being imprisoned, losing their jobs, and more. More recently, the fear of terrorism has had dramatic effects on many functions of the USA's society. The fear of terrorism on airplanes has created a nuisance for people who are traveling because of the heightened security controls at airports. It has also made entrances to many public buildings have more extensive security checks.

In many animal species, the male dominates a group of females. In many cultures such as and perhaps most distinct in Islam, but also in Christianity, it is common that men dominate women in various ways. This domination is done mainly through the creation of fear of repercussions.

"Neither a man nor a crowd nor a nation can be trusted to act humanely or to think sanely under the influence of great fear." (Bertrand Russell: See *http://theunboundedspirit.com/the-weapon-of-fear-how-they-use-fear-to-manipulate-you.)*

Fear is used to get people who have no symptoms to go and get extensive regular checkups. Excessive checkups have been shown to provide little benefit to people who do not have symptoms of a disease or hereditary high likelihood for acquiring specific diseases [10]. Many medication and surgical treatments that have been promoted by inducing fear have been shown to have little or no benefit to a person, and sometimes have an adverse effect.

Inducing fear can change the stock market, usually for the worse, faster than anything else and to a greater extent than most factual matters. Unsubstantiated optimism can do the same, thus causing people to have less fear than what would have been justified or found beneficial. Rumors are more effective in moving financial markets than facts. Belief has a tremendous effect on the prices of goods and the markets for oil, gas, metals, and other goods.

Currencies may be traded wildly because of induced fear, and fear plays an essential role in the values of currencies and exchange rates. Fear of political unrest moves financial markets erratically and often cause massive swings. Fear can change the price of many vital commodities in a country, such as fuel prices, and this fear can have a significant impact on a country and the people.

Examples of human-made fears that had catastrophic consequences

Incorrect information or the misunderstanding of circumstances is often the basis of many human-made scares. Here is an example of how a single person was able to cause great fear in many people and, because of that fear, cause death and severe illness to many children.

An article from 1998 [11] in the prestigious medical journal, The Lancet, a British surgeon, Mr. Andrew Wakefield, claimed that he found vaccines or the additives in the vaccines that could cause autism. He based his claim on his own study of 12 children with autism who had been vaccinated with the measles, mumps, and rubella (MMR). The article concluded that vaccination might be the cause of autism.

The author, Mr. Andrew Wakefield, a surgeon specializing in gastrointestinal surgery suggested that a new variant of autism that was associated with intestinal inflammation could explain the association between vaccinations and autism. He proposed the administration of the MMR vaccine as a possible cause of the inflammatory process and autism.

The information spread to the entire World through news media outlets that presented the information to the general public. They stated that the vaccination usually given to young children might cause autism. It created a deep fear in parents about having their children vaccinated. It deprived many children of essential protection against diseases that could cause death and severe damage to the brain and through that neglect many children died unnecessarily, and others suffered and still are suffering various degrees of brain damage.

It was later found rigorously performed studies that there was no evidence of a relationship between vaccinations and autism. Further studies were designed primarily for that purpose and were. In 2004, it became evident from many studies that there was no evidence at all for such a relationship between vaccinations and autism. Brian Deer, an investigative journalist, conducted an accurate investigation that revealed how the Wakefield research had many faults and was performed with a predominantly economic motive or objective. The story was published in the British Medical Journal (BMJ) [12].

In 1996, Mr. Wakefield was approached by lawyers representing an anti-vaccine lobby. In addition to the study being from a weak and biased scientific point of view, it turned out that the result Mr. Wakefield reported was, in fact, fabricated for economic winnings by lawyers who represented parents in medical lawsuits. The lawyers were aimed at suing drug companies who manufactured vaccines. They supported Wakefield's research to enrich themselves and their clients. As a result, Andrew Wakefield left England and as of 2012, he worked in Texas for anti-vaccine lobbying groups.

The original paper by Wakefield published in the prestigious journal, The Lancet, was retracted, and Wakefield's medical license in England to practice medicine was retracted. In 2010, Wakefield was expelled from the General Medical Council in the United Kingdom.

Retracting Dr. Wakefield's medical license seems like a slap on the wrist for a person who caused fear among so many parents and directly caused many children's deaths and others to be disabled for a lifetime. The ill effect of Dr. Wakefield's action still lives on now more than ten years after it was disclosed that there was no ground at all for the claim that vaccinations could cause autism. This is yet another reminder that people's belief is difficult to correct and that it takes a very long time to correct an erroneous belief.

This terrible story and the events that followed is an example of how one person can cause a great amount of fear in many people and directly be the cause of death and severe illness of many children. It is evident that Andrew Wakefield had help in causing this unjustified fear by journalists who uncritically cited parts of an article in the Lancet, excluding the fact that only 12 patients were studied. The articles published in the general press or television coverage also failed to mention that the particular form of autism occurred simultaneously with intestinal inflammation.

While physicians and surgeons can cause fear in people one at a time the story of Mr. Wakefield is an example of how, journalists can cause fear in many people at the same time if they are not careful with the way they translate the results of scientific studies. Journalists often act as a "filter" or interpreters of the results of scientific studies. There can be many reasons for such "filtering" and interpretation of scientific data. The primary and ultimate purpose should be to make the results of complex studies understandable for people who do not have a scientific background. It is well known that leaving out or tweaking data to fit a hypothesis that is beneficial or exciting can make results of studies look more sensational.

With time, science becomes more and more complicated, and it is understandable that journalists are not able to explain to a general audience or readership what they do not understand themselves. Many scientists are good writers, and it would be better to let these scientists explain scientific matters in the news media.

There are many other examples where fear has been stirred up about matters that have no possibilities to affect people of the general public. One recent example is the panic from the Ebola virus that evolved from a person with Ebola who came to the USA. Public polls showed that as many as 40% of the general public were afraid of contacting the disease. This is naturally completely unrealistic. At the same time, many people avoided getting the vaccination for influenza, which kills as many 20,000 people in the USA every year on average (and 80,000 died of influenza in 2018). This is a stark contrast in the probability rate of contracting certain illnesses and the level of fear evoked.

People's response to fear varies

Peoples' response to fear varies between individuals, and it depends on many factors, including genetics, cognition, and learning. Different people describe fear in a variety of ways. As previously stated, it is similar to the Indian fable about the six blind men who were describing an elephant. There is no doubt that a person's education, occupation, and professional background will affect a person's view of fear. A psychologist, a psychiatrist, a neuroscientist, a social worker, an economist, and a politician all will have distinct views on fear, but their views will be different and distinct.

For better or for worse, fear provides attention to matters that may cause bodily injury, and it makes people aware of symptoms of disorders of various kinds. Fear signals the need to take appropriate action in response to risks of bodily injury. The main response being to either run away or fight. Ideally, the degree of fear should be related to the risk of the occurrence of a harmful event and the amount of harm that may result if the harmful event in question would occur.

However, the fear many people experience is rarely related to the actual risk of harm. The experience of fear controls many of a person's reactions to both real and hypothetical situations. It can make people avoid situations they assume are dangerous and thus cause a high degree of fear, and it can avoid appropriate reactions to situations that have a high degree of risk of harm.

For example, a comparison of the fear and risk of two very much discussed situations, such as terrorism and a common disease such as influenza, illustrates a lack of correlation between a person's degree of fear and the risk of being affected by a dangerous event.

A rational person should regard terrorism in the USA to be a much smaller risk to a person than common influenza. Terrorist activity has a minimal risk of affecting a person's life in the USA because it has in the past caused very few bodily injuries, and deaths and influenza are regarded as a more significant risk because it has caused more death to people living in the USA (less than 5 deaths annually versus an average of 20,000 deaths for influenza).

Many severe risks that a person may experience can be reduced by appropriate actions, such as reducing the risk of illness and death from infectious diseases through vaccination. However, other risks are beyond control by a person such as the effect of terrorism. Even though these facts are well-known, many people pay more attention to the risk of terrorism.

The action people take in response to fear also varies among individuals. Their responses are often out of proportion due to the implications of the fear, and it may be vastly exaggerated such as avoiding traveling by commercial airlines or a person who is ignoring real dangers. A person's response to fear can be a confrontation with the threat, or it can be a catastrophic reaction, fearing that bad things will be worse in the future. Often people seem to take too much action in their action to matters that does not pose a danger but there are also many situations where people take less action than justified. An example is not getting vaccinated such as for the influenza.

Fear plays a part in many other essential roles in making people obey rules and laws by someone inducing fear of the consequences of not complying. Similar methods are also practiced extensively in modern societies and between people such as in marriages or parents' treatment of their children. The educational system, at different levels, makes use of fear to get children or students to pay attention and to make efforts to learn. It was mentioned above that people can use fear for their benefit such as for getting personal power or influence or for personal enrichment. This is just another way people differ in their way of using fear and responding to fear.

Many things that previously posed a considerable risk of death and disability now possess little or strongly reduced risk if appropriate actions are taken. In the past, the incidence of many diseases that caused many deaths can now be avoided by vaccinations, and some can be cured all together. Transportation has been improved, and transport by commercial airlines has reached a very high level of safety. Areas where the level of risk have increased are forms of criminality where the risk of harm is brought about by an action of other people. The amount of money spent by the US Government is not in proportion to the risks of occurrence and the seriousness of the consequences due to risks such as terrorism, common diseases, or criminality.

Factors that can influence the perception of fear

The outrage factors

A person's perception of different kinds of risks depends on objective factors such as the statistical chance that an untoward event may occur besides irrational factors. The outrage factor is an irrational factor, and the outrage factor determines the attention a certain kind of event will receive from the news media. If a story is given this kind of attention, the intense coverage is what scares people. Sandman has collectively called these additional factors the "outrage". Sandman assumes that the risk of an untoward happing contains two factors, one of which he calls the real risk of the hazard, and one or more factors that he lumps together under the name of the outrage factor [13].

The outrage factor is a group of irrational factors, that represents a person's overstatement or understatement of risk. This means that people's perception of a risk not only dependent on the likelihood that it will occur and the consequences if it occurs, but it also depends on various irrational factors. This irrational factor has also been called an "imaginary" component because outrage is associated with an emotional reaction of unspecific factors to specific risks.

Generally, the outrage factor plays a bigger role in people's perceptions and reactions to risks than does hard scientific information. Experience may either increase or decrease this unreal factor. When the unreal factor has a low value a person's perception of a risk is rational, which is rare.

These outrage factors can both increase and decrease peoples' perception of a particular risk. If a person perceives a specific risk to be higher than the real risk, then this means that the imaginary component has added to a real risk. The outrage component may be a result of biased information about accidents, or it may be the news media's promotion of risks in general. For example, people seem to be more worried about accidents in nuclear power plants than the pollution of mercury caused by traditional power plants fueled by fossil fuel.

If a person perceives a risk to be lower than the real risk, then the imaginary component has subtracted from the real risk. The imaginary component, thus represents the difference between the real risk and how a particular person perceives the risk.

Different kinds of fear have different outrage factors.

The real part of risks can also be regarded as having two parts, one over which a person has control, and one where a person has no control. According to Bernstein [14], "The essence of risk management lies in maximizing the areas where we have some control over the outcome while minimizing the areas where we have absolutely no control over the outcome and the linkage between effect and cause is hidden from us". However, people's impression over what they have control is often exaggerated. People understand they have no control of a scheduled airline, but they believe they can control the automobile they are driving. The latter is partly true. People can avoid driving when accidents are more likely to occur, such as between 2 and 4 AM, and people can exercise defensive driving. All that will reduce the risk of an accident but not eliminate it.

An example of where the outrage factor may change in the future is related to the risk of terrorism, which at present is very low in the USA. However, there may be a noticeable risk if, for instance, terrorists would get their hands-on nuclear material or a dangerous poison of some kind. There is a level of uncertainty that adds to the outrage factor. Automobile accidents or casualties at our hospitals does not change dramatically from year to year, but the casualties of terrorism can possibly be very large even though there may be a very small likelihood that will occur. This is one reason why terrorism is associated with a very high outrage factor.

This means that people place an unrealistic evaluation of the risk of bodily harm and death on many forms of harm. The value depends on many factors. The total number of people who are being killed plays a minor role, but the circumstances play a large role. There is also pronounced time decay. The first time an event occurs, it gets the most attention, if it repeats it gets less attention.

Important decisions that may affect the risk of death and bodily harm are sometimes made while an individual is under the influence of substances that affect judgment. An example of a risk that is affected by an imaginary factor of that is related to driving while under the influence of alcohol or other drugs that affect a person's judgment the effect on the brain and consequently on the decision making of a person from an additional factor that may be called the outrage factor. Decisions on whether or not to drive while under the influence of alcohol are usually made after that the person has become intoxicated. Many people who drive while drunk would not have elected to drive in that condition if alcohol had not impaired their judgment. Another example of an outrage factor influences a person in decision making is related to sexually transmitted diseases. Decisions regarding sexual intercourse made while a person is intoxicated may be affected by a poor judgment resulting from intoxication and by that expose a person to a higher risk of sexually transmitted diseases.

Why do some events cause outrage while others do not?

Different kinds of risks can have very different outrage factors. In general, events that cause many casualties at one time have much higher outrage factors than events that cause a steady stream of casualties. Airline accidents have much greater outrage factors than automobile accidents despite the fact that automobile accidents cause many more casualties.

Small risks may cause significant emotional responses, while activities that take high tolls of lives and cause many injuries cause outrage. Accidents with many injured and killed people in the same accident attract much more attention than accidents in which few people are killed at one time, despite the average annual deaths being small compared with the annual deaths and injury tolls of other accidents. Crashes of airplanes of scheduled airlines that cause an average number of deaths that is less than 100 per year in the United States, but MVAs (motor vehicle accidents) per year cause approximately 40,000 deaths. The safest way of traveling is by commercial airlines and by busses. Nevertheless, airline accidents receive extensive coverage by the news media and evoke endless discussions about how to improve the safety of commercial airplanes.

Acts of terrorism always cause outrage possibly because it is done on purpose. One person harming others for the sake of a goal or belief is dramatic. Deaths caused by automobile accidents are called "accidents," indicating that they are unavoidable, and the killing or injuring was made without a purpose. That is the reason that such events do not result in "public outrage".

The fact that the risk of automobile accidents is much higher when a driver is under the influence of alcohol or drug is not perceived in the same way as killing on purpose is. It is, however, odd that "outrage" is immediately triggered if a bus driver, a train engineer, or an airplane pilot is found to have been under the influence of alcohol while working. Thousands of intoxicated automobile drivers travel on our roads every day, but that does not cause significant outrage, and most people are not willing to do anything serious about the problem.

Ultimately, a particular behavior seems to be accepted in one situation but not in another. Perhaps this is because affecting drunk driving practices in private automobiles is more difficult than it is to affect drunk driving of bus drivers and train engineers. One reason that traveling by bus is so much safer than automobile travel is probably that intoxicated bus drivers are few. The consequence of an intoxicated bus driver is much worse than an ordinary person; a bus driver loses his/her job if driving while intoxicated in addition to legal troubles. Another reason that we may believe that is because in a person's own automobile the person should be free to use it however and whenever he or she wishes.

The spreading of fear and the focus on small and imaginary risks is a distraction from real risks, and it causes unnecessary worry which decreases the quality of life. We may sometimes forget that mental health is also essential. Issuing warnings when no real dangers exist not only creates unnecessary fear, but it can also have the effect that nobody will pay attention when a real emergency occurs ("crying wolf too often"). Naturally, the legal system has an interest in fear because the results of that fear provide work for lawyers and others in the system.

It is often reported that an "unusual" accumulation of specific rare diseases occurred at a certain geographical location and people will search for a cause for that. However, when such reported "clusters" were investigated, it has almost always turned out to be the result of normal statistical variations. The nature of chance is so that sometimes a pattern will appear that looks as something caused it, but we know that correlation does not imply causation.

How to decide what to worry about?

Decisions are often based on the perception of the risk of harm and not on the *actual* risk. An incorrect perception of risks often controls or directs a person's choice. Example: A person may choose between flying (commercial) or driving a personal automobile for safety reasons disregarding the fact that it is far riskier to drive than fly.

Too much fear occupies a person unnecessarily and reduces the quality of life while wearing unnecessarily on the body. Too little fear makes people embark on reckless activities such as driving recklessly, wild speculations on the stock market, and having sex with people they do not know.

The meaning of safe, dangerous and risk

The words such as "safe", "dangerous", and "risk" are used often. For example, one would hear that exposure to tobacco smoke, ionized radiation, and the sun's rays poses a risk. These statements are of course true, but many other aspects of daily life also pose risks. It is the degree of risk that is important. It is often stated that secondhand smoke posed a danger, and it does so even in very small amounts and for brief times. This is of course true but so many other activities are dangerous and what is important is the size of the risk. Risk from smoking cigarettes is obviously greater than being exposed to secondhand cigarette smoke. Being exposed for a short period must carry less risk than being exposed for a long time. These are examples of misleading use of the word "danger". Information that does not include the size of a risk can do more damage than good by spreading unspecific fear. The use of the word "risk" without specifying the degree and likelihood of occurrence can be equally misleading. For example, the radiation that poses the greatest risk for most people is not nuclear but comes from the sun's rays. Despite that, most people fear radioactivity much more than exposure to the sun.

Risk from vaccinations and from exposure to magnetic fields have been debated vigorously over many years, one group of people saying it is dangerous, another saying it is safe (or not dangerous). Both groups use the same data; just the interpretation is different and more important, they use the words "dangerous" and "safe" to describe the same data, instead of providing numbers to describe the risk.

Fear is based on a belief

A person's perception of danger mostly depends on belief and often ignoring the facts. Ideally, belief should be based on information about facts, but it is often based on a person's imagination. The information about risks may be incorrect and incorrect information may be spread because the person who spread the information may not know its level of correctness, or incorrect information may be spread for a purpose.

Unsupported and unfounded fear may cause harm because fear in itself is harmful. This kind of fear may make a person make choices that are not beneficial for the person. Unfounded fear may be the most important form of fear.

Truth

A person is often afraid of getting sick, and many people are unjustifiably afraid of getting a specific disease, but it is nearly impossible to predict what disease a person will get. People often get a different disease different than the one they were afraid of getting. It is, therefore, a waste of time and energy to worry about a specific disease.

The nature of fear and a person's perception of fear

Adolphs [5] has asked a series of important questions regarding fear, its nature, and how it is perceived. He wondered if we are discovering 'fear' through objective scientific investigation. Do we use similar methods as in other branches of science such as physics? Alternatively, do we regard fear as a psychologically constructed category, and, of course, does that not make it less biological? Adolphs states in one of his articles [5] that the answer to a person's worry depends on assuming patterns seen by scientists are also patterns seen by evolution.

The reason that our definition of fear is not singular is because it is anchored in both behaviors and stimuli. Specific sets of stimuli and behaviors simultaneously vary; if they did not, we would never be able to attribute fear to other people or animals, but we can [5].

The degree of fear that a person may experience, from a specific harmful or adverse event or phenomenon, should ideally only depend on two factors: (1) the likelihood that the harm in question would occur and (2) the severity of the expected consequences. Instead, however, the fear a person perceives is related to the person's subjective evaluation of the degree of risk and his/her assessment of the consequences of the risk. Most of the fear that is experienced under different circumstances is exaggerated, and in many instances, the fear a person experiences is gravely exaggerated. Examples, in the USA, would be the fear of death from terrorist actions or flying on commercial airlines. Some forms of threat are, however, ignored, such as that of some diseases or of health risks related to what a person eats, or lifestyle in general and what preventive actions a person takes such as vaccination for infectious diseases.

In the ideal world, the degree of fear a person perceives would be directly related to the size of the threat and the amount of danger or the importance a person ascribes to the threat. However, in the real world, the fear a person perceives may be different from the real threat. Belief is based on many factors, and only some of which are based on objective information such as statistics on the likelihood of occurrence of a particularly dangerous event.

A person's fear of a specific matter or event can be modified by cognition, and it varies widely among individuals. The magnitude of a person's, harmful or not, is affected by factors such as memory, history, and a person's temperament.

The fear a person perceives often depends on irrational factors in addition to a person's social and professional status, power, wealth, and security. Therefore, a person's perception that something harmful may happen is often unrelated to the real likelihood of the occurrence of an event and the severity of the consequences if it occurs.

Many people feel afraid of things that pose little risk of harm to a person. The cause of that is because people base their fear on a belief instead of facts. News media outlets, in the USA particularly, are not helpful in correcting people's understanding regarding what poses a risk of harm to a person. In fact, news media outlets often make many people afraid of things that have little risk to a person. Specifically, television has a strong influence on what a person may fear and because the news media reports incidences of dramatic events that cause several deaths each while trivial events cause the majority of deaths and injuries, one at a time. Thereby, news media outlets induce unnecessary fear.

News media outlets provide a filtered version of the information. One form of filtering is related to time. Only the new matters and events are reported. This is why it is called "news media"; it is a money-making industry. Another form of filtering is induced to satisfy people's preferences for exiting matters. That is why the news media present reports on spectacular matters such as airplane accidents. Similarly, the news media outlets do not report on some important risks because it is not exciting, such as falling as a cause of illness and death or infectious diseases and the value of vaccinations. Only what is supposed to catch people's attention is reported by the news media because the news media wants to have as many people as possible listening or watching their program since that is what generates advertising money.

For the same reason, big news media outlets focused on a person who died from the Ebola disease and ignored many other much more important diseases such as influenza that takes an average of 20,000 lives annually in the US (in the year 2017, 80,000 people died from the influence). The risk of dying from influenza can be greatly reduced by vaccination.

What can fear cause?
Fear affects many body functions, and it is particularly associated with several reactions from the autonomic nervous system. The autonomic nervous system is activated by fear. Threat perception activates the sympathetic part of the autonomic nervous system and deactivates the parasympathetic part, leading to increased release of the catecholamines epinephrine and norepinephrine.

The fear of different strengths elicits different reactions of the nervous system:

1. Intense fear can elicit a flight or fight reaction, with much elevated sympathetic and low parasympathetic activity, and difficulty in performing mental tasks.

2. Moderate fear causes elevated sympathetic activity and decreased parasympathetic activity.

3. Slight fear causes sympathetic activity to be somewhat elevated and lower than the usual parasympathetic activity that causes muscles to be tense. It can cause a distraction that may make it challenging to do intellectual work.

4. Concerns cause slightly elevated sympathetic activity and a moderate decrease in parasympathetic activity.

5. Freedom of fear causes normal sympathetic activity, regular parasympathetic activity, muscles are relaxed, and rational ability to concentrate on mental work.

The autonomic nervous system is activated to prepare the body to face an emergency, the flight and fight reaction being one such reaction. Fear may also result in fainting (vasovagal syncope), which means that the vagus nerve is activated which causes low blood pressure. Extreme fear can cause the "freezing" reaction making it impossible to move, thus affecting the parts of the brain that control muscles (motor systems). The unpleasant features of fear are mainly mediated through the amygdala. Fear and pain can also give a feeling of pleasure, and this feeling is probably a result of connections to the award system of the brain such as the nucleus accumbens.

The autonomic nervous system regulates and controls many of the body's unconscious functions such as blood pressure and the activity of the digestive and reproductive systems. The autonomic nervous system is continuously active in maintaining homeostasis *i.e.* maintaining stability and functioning within a normal range of a stable equilibrium in systems such as different body systems where there are many complex interactions between different systems.

The autonomic nervous system consists of the sympathetic and the parasympathetic systems. These two parts have the opposite effect on many vital systems such as blood pressure, heart rate, and the digestive system.

The parasympathetic nervous system which stimulates the body's ability to "rest-and-digest" or feed and breed is described as being complementary to the sympathetic nervous system. The parasympathetic nervous system activates the "housekeeping" systems such as the digestive system. The activity of the parasympathetic nervous system is reduced in a situation of fear. For example, there are better times to digest food than when the body is at risk. Threat perception leads to activation of the hypothalamic-pituitary-adrenal [15] axis leading to increased release of the glucocorticoid hormone, cortisol, from the adrenal glands [16].

This is one of the ways that fear can suppress the immune system. The effect of fear on the autonomic nervous system also has a suppressing effect on the immune system. This includes increasing inflammatory processes with all the subsequent effects on diseases of various kinds (for details see Chapter 4).

Fear puts the body in a condition that is beneficial to fight by redirecting nutrients to muscles from the skin and internal organs except for the heart. This means that many structures of the brain are indirectly involved in fear reactions. Being afraid can sometimes cause "freezing" which prevents, or at least makes it difficult, a person from moving or running away. This is a protective action because it may be safer in a dangerous condition to not move. Freezing is more pronounced in some animals such as the rat, which easily freezes.

The autonomic nervous system also controls functions related to reproduction, and it controls many glands and thereby controls the secretion of hormones and other substances such as adrenaline.

Different kinds of fear that can affect a person

There are many kinds of fear, and some forms of fear have no or very little likelihood to materialize. Some forms of fear and anxiety harm a person not only by reducing the quality of life, but because a person who has strong fear and intense anxiety may often appear like a loser in general. The same is the case for people who excessively pessimistic contrary to people who are optimists. These matters are discussed in more detail in chapter 4 where the influence on people from fear brokers is also covered.

People's belief can control many of a person's reactions

Fear is, to a great extent based, on belief and not on facts. A person's belief controls his/her reaction to the risk of injury to a greater extent than facts. Fear is beneficial like pain that warns about the risks to a person's body, but the fear is often misdirected. For example, people are afraid of flying but do not fear driving in a personal automobile.

Many decisions are based on what a person believes rather than facts, such as research results or estimations based on facts. People tend to rely much more on what they believe than what is reality or what is challenging to change a person's belief or opinion. Solutions to significant problems such as how to best take care of the human body are often based on what a person believes, and actions based on a person's belief are often inferior to actions based of science. Belief is also often a significant obstacle in introducing new results and methods of handling diseases that are based on science.

There are many examples of how belief can be a hurdle in rationally handling diseases and other adverse effects on the human body. Also, convincing people that prevention, decreasing the risk of acquiring a disease, is often far more effective than treating the disease when it occurs.

It is difficult to change a person's belief

It is well-known that it is difficult to change a person's belief. There are several reasons for that. Why is it so difficult to change a person's belief regarding what poses a risk and what does not? Accepting what is generally accepted as the truth or what is shown by the news media results in a pleasant feeling because what is supported by public or conventional findings elicits the release of dopamine in the brain activating the reward system.

On the other hand, when a person does not accept what is shown to be a publicly accepted hypothesis, the amygdala and the same brain circuits as those triggered in fear become activated [17]. That can explain why many people resist changing their beliefs. Many people focus on information that supports the hypotheses because they are used to accepting and ignoring information that contradicts or does not conform to their initial analysis. This is also applicable to medicine where there are many examples of people who get in trouble because they believe in the ideas or knowledge, they "know for sure" but which are not true. For example, many people rely on the treatment of diseases and ignore preventions while the reality is different. It took ten years before the finding that stomach ulcers are often caused by bacteria was accepted; it was always previously believed that a stomach (gastric) ulcer was caused by stress or nervousness, thus originating from signals from the brain.

The reaction to confirming data versus opposing data is different. Confirming data causes a dopamine rush, similar to the one we get if we eat chocolate, have sex, or fall in love. The changing of a person's belief can cause fear. Opposing the generally accepted hypothesis or belief is much more difficult and causes activation of brain structures including the amygdala causing fear.

People often focus on information that supports the data or hypotheses and ignore everything that contradicts or does not conform to our initial analysis. (N. Hertz, "Eyes Wide Open: How to Make Smart Decisions in a Confusing World" [18]) (Gorman, Sara E.; Gorman, Jack M. Denying to the Grave: Why We Ignore the Facts That Will Save Us. Oxford University Press. [17])

Confirmation bias as good as eating a candy bar

Using confirmation bias to make decisions may actually make us feel good in the way that people experience the positive effects of alcohol or opiates. The "reward" pathway in the brain that runs from a structure in the brainstem, the ventral tegmental area (VTA) to regions deep in the lower part of the brain, the nucleus accumbens (NA, NAc, or NAcc) is activated. This pathway releases dopamine whenever a person (or almost any animal) anticipates or experiences a rewarding or pleasurable stimulus.

Studies have shown that it is fundamentally challenging to alter commonly accepted opinions and opposing conventionally accepted ways to handle matters, and these situations activate brain structures that are usually involved when a person experiences fear including the amygdala [18] [17]. For example, it seems to be the fact that diseases should be treated, but it meets resistance when trying to convince people that it is better to make efforts to prevent diseases than treat diseases.

In an experiment by Gregory Berns [19], it was found that compliance with social rules or laws was associated with a decrease in activation of the frontal lobe and an increase in activity in the parietal and occipital lobes. The parietal and occipital regions of the brain are where perceptions are formed. When participants in this experiment acted independently of the group, there was more activation in the amygdala.

Resisting the participants' group is a frightening proposition that must hold the promise of some reward to be tenable. A decrease in biased thinking is associated with a decrease in amygdala activity, while changing one's mind or going against the decision of others is associated with activation of the same circuits in the brain as those that are usually activated in connection with fear [17].

Saying yes is easy, saying no is difficult

Similarly, the confirmation of data or hypotheses is easier than opposing the established hypotheses, it is also easier to say yes than to say no. The following are excerpts from a newly published speech by the former seventh director of the FBI, James B. Comey, Esq that discusses saying yes versus saying no. In his address to the National Security Agency (NSA) in May of 2005 that was published in the summer issue of the Green Bag, a legal affairs journal, James Comey said, "It is the job of a good lawyer to say 'yes.' It is as much the job of a good lawyer to say 'no.' "No" is much, much harder. "No" must be spoken into a storm of crisis, with loud voices all around, with lives hanging in the balance. "No" is often the undoing of a career. Moreover, often, "no" must be spoken in competition with the voices of other lawyers who do not dare to echo it."

This also applies to other professions. For a scientist, it is easier to accept current hypotheses than to reject current hypotheses. However, rejecting current hypotheses often brings new concepts forward, and questioning and rejecting old hypotheses are among the most effective ways to promote progress.

Similar reasoning is valid in medical science and, in particular, in clinical medicine where there are many reasons for accepting current hypotheses and theories about treatment and, there has been much resistance in attempts of rejecting old kinds of treatments. An example is what was mentioned above about the cause and subsequent treatment of the common disease, gastric ulcers. The new and much more effective treatment that was developed from these new finding has changed millions of people' life.

The situation about saying yes or no relates to the discussion about the confirmation bias. Saying yes is equivalent to the confirmation of a generally accepted hypothesis while saying no causes the same neural activation as opposing a generally accepted hypothesis. This means that the dopamine pleasure system of the brain is activated when saying yes, while saying no activate the same neural circuits as are activated in response to fear, thus an unpleasant reaction.

Cognitive mechanisms of optimism and pessimism

Optimism and pessimism are associated with distinct perception and cognitive modes. The principal differences between optimism and pessimism are:

a) Selective attention and information processing.

b) A belief, or lack thereof, that one has the power to influence relevant situations, events, and relationships (i.e., the locus of control).

c) The general schema one holds for interpreting personal events varies (i.e., attribution style). (Hecht 2013).

Anxiety

Anxieties are forms of worry or fear from real or imaginary threats. Everybody has experienced anxiety at some point in his or her lives; some experience anxiety more frequently than others, and different people are affected differently. Anxiety disorders can cause extreme distress that affects a person's ability to live a normal life. The term anxiety includes post-traumatic stress disorder (PTSD with an estimated prevalence of 6.8%), phobias (now called social anxiety disorder or SAD with a prevalence of 12.1%), and panic disorders (prevalence of 4.7%).

Specific phobias (prevalence of 12.5%) are marked by a persistent, excessive fear of a specific object or situation (classified as animal type, natural environment type, blood-injection-injury type, situational type, or another type [20] (DSM IV., Washington (DC): American Psychiatric Association; 2000).

Signs of anxiety

While fear is directed to specific causes such as risks of untoward events, anxiety is a general and unspecified reaction to unwanted matters. Anxieties are perhaps best described as the exaggerated and often irrational worry about matters that pose little risk. Anxiety is a more general feeling of worry or unease such as from the expectation of a future event that is perceived to involve a danger. For example, it is common to be anxious about a visit to the hospital and awaiting the results of tests. Any medical examination can cause anxiety in many people. It is also common to be anxious for upcoming exams or a job interview, or for the risk of being laid off or not getting an expected promotion.

While the term fear is related to the possible occurrence of specific events that are dangerous to a person, the term anxiety is used to describe phenomena that have similarities with fear but not related to the occurrence of any adverse events. Anxiety is closely related to threats that are perceived to be uncontrollable or unavoidable [21]. Anxiety can be described as a form of general and unspecific fear. Many neurologists and psychiatrists regard some forms of anxiety to be a disease. Tremendous effort and capital have been spent on treating anxiety.

Anxiety has many adverse signs, including nervousness, worry, or uneasiness in anticipation of an imminent event that involves an uncertain outcome. Anxiety varies among people, and it is related to a person's personality and tends to change less over time than fear. Some people are worried about certain things while other people worry about a multitude of things.

Anxious persons also show a tendency to engage in cognitive-behavioral avoidance. This avoidance limits their ability to challenge inappropriate threat perception, confront and resolve threatening situations, and reshape expectations for the future. The ultimate result of this process is a failure to achieve a resolution of perceived threats, resulting in sustained threat perception. (From [16].)

Anxiety has many similarities with fear, but they are in fact two independent entities with their own definitions. Anxiety is characterized as a mental state that arises from general and non-specific stimuli that are perceived as being potentially threatening in the future. This perception often results in an apprehensive mood accompanied by an increased level of arousal and vigilance which, when taken to an extreme, persist for an extended time.

Fear is evoked by specific stimuli or events such as active defensive responses. Fear can be the response to a specific cue, such as seeing a snake. When the specific stimulus is no longer present fear may gradually diminish, and the resulting fear may disappear altogether. The anxious person reacts stronger to fearful situations and worries more than the less anxious person. That is one example of how fear and anxiety are not independent entities.

Anxiety involves anticipations of when danger will occur based on the form and strength of the threat.

1. Fear is a biologically adaptive physiological and behavioral response to the actual or anticipated occurrence of an explicit, threatening stimulus. Anxiety crucially involves uncertainty as to the expectancy of threat and is triggered by less explicit or more generalized cues.

2. Anxiety is characterized by a more diffuse state of distress, with symptoms of hyperarousal and worry.

3. The mass media outlets bring news of natural disasters, potential pandemics, terrorist atrocities, and violent crime straight into our homes.

4. Perhaps it is not surprising that nearly one in four of us will experience a clinical level of anxiety within our lifetimes.

5. It is interesting why only some persons experience excessive fear, worry, and disruption to an everyday function that characterizes clinical anxiety.

Some anxiety disorders have been associated with stress of various kinds. Many studies have provided substantial evidence that stress affecting epigenetics plays a role in creating diseases of anxiety, especially regarding the effect on the HPA axis.

Anxiety has often been regarded as having similarities to fear, but anxiety is not identical to fear. Meriam-Webster's dictionary defines anxiety as fear or nervousness about what might happen; a desperate feeling of wanting to do something. Anxiety may affect the perception of fear. This means that what extreme fear a stimulus can cause is different in an anxious person compared with a person that is not anxious [16]. The emotional reactions that fear causes in an anxious person result in a different reaction to cognitive problems such as failure to resolve problems that the person who is not anxious usually can solve [16].

Anxiety is a long-lasting symptom that is not specific to overt cues. Fear can be a part of anxiety disorders with symptoms such as, but not limited to, apprehension or general uneasiness. Anxiety is a term that is often used to describe a person's emotional reaction.

It is often attempted to study anxiety in laboratory rodents, but it is questionable how well the behavior of an animal mimics human anxiety. Animal models of fear may be better at mimicking human fear, but it is still questionable how accurate the animal model of a rat can mimic human conditions of fear [22]. The level of anxiety varies widely among people. Situations that cause anxiety in one person may not cause another person to become anxious. A person's fear can be affected in many different ways, whereas anxiety is a trait that is very different in different people. Also, anxiety is related to a person's personality, which remains more or less constant during a person's entire life, and it is difficult to change. Intense anxiety, chronic or acute onset, is often regarded to be a disorder. Anxiety can have many causes, including genetic or environmental stimuli or circumstances, but it also occurs without any known cause.

An anxious person often has cognitive biases toward threatening information, which leads them to detect threatening stimuli (e.g., angry faces or predatory animals) more quickly than non-anxious individuals. Anxious people appraise both ambiguous and threatening stimuli as more threatening than normal. To reiterate, anxious people also often show a tendency to engage in cognitive-behavioral avoidance, which limits their ability to challenge inappropriate threat perception, confront and resolve threatening situations, and reshape expectations for the future. The ultimate result of this process is a failure to achieve resolution of perceived threats, resulting in a sustained threat perception [16].

People with a high level of anxiety respond differently to threats than people who are not anxious (Figure 1.1.)

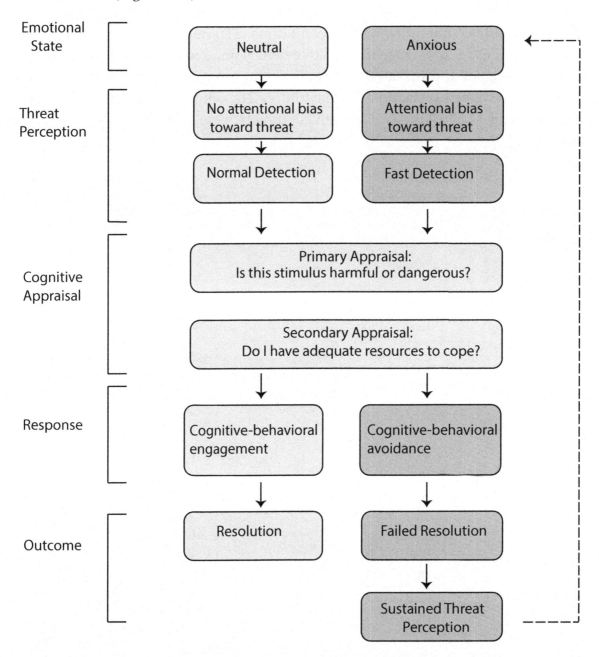

Figure 1.1 A broad overview of cognitive-behavioral responses to perceived threats in anxious and non-anxious individuals (Based on O'Donovan, et al 2013 [16].) (Artwork by Liliana Cabrera).

Anxiety affects many aspects of life, and it is associated with higher risks of acquiring diseases of various kinds as well as affecting the aging process. The anxiety experienced by one person may be different from the anxiety experienced by another person in the exact same situation. The anxiety experienced by one person may be the same that another person experiences from a completely different threat.

Beneficial effect of anxiety

A certain degree of anxiety may have a beneficial effect on performance, and anxiety has an enhancing effect on memory that may be another beneficial effect. An anxious person may think the worst of situations, which can be good at certain times and in certain situations, but it can unnecessarily occupy a person's mental abilities and cognitive economy.

Anxious people may have fewer accidents because they are frequently concerned with their surroundings. Anxious people may think things over to a greater extent than people with a lesser degree of anxiety, but in fact, anxiety is essential for many kinds of human activities such as creativity. Kierkegaard, the Danish philosopher, has written about the role of anxiety and stated that anxiety powers creativity. Anxiety may take the form of a disease when abnormally strong, and such maladaptive emotions can benefit from treatments such as medical and behavioral therapies.

Anxiety helps maintain a state of readiness. It facilitates threat processing and defensive responding, but it also prompts cognitive changes. To clarify, anxious individuals show a cognitive bias towards threatening information. This cognitive bias in anxious people, which leads them to detect threatening stimuli (e.g., dangerous drives or illness prone situations) more quickly than non-anxious individuals, gives them an advantage or disadvantage when appraising both ambiguous and threatening stimuli as more threatening.

Harmful effects of anxiety

There are many adverse effects of anxiety. Anxiety may affect bodily functions such as the autonomic nervous system and endocrine functions. Anxiety may be an unpleasant emotion, and it may affect a person's cognitive abilities. Anxiety and depression are often combined in psychiatric diagnoses.

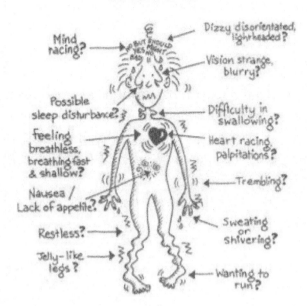

Figure 1.2 Anxiety affects the function of many systems of the body.

Anxiety is often regarded to be an undesirable emotion; in fact, in modern society, anxiety is one of the most common conditions to be diagnosed and treated. Generalized anxiety disorder is defined in the Diagnostic and Statistical Manuel (DSM, currently DSM-5) of the American Psychiatric Association as:

A. Excessive anxiety and worry (apprehensive expectation), occurring more days than not for at least 6 months, about a number of events or activities (such as work or school performance).

B. The individual finds it difficult to control the worry.

C. The anxiety and worry are associated with three (or more) of the following six symptoms (with at least some symptoms having been present for more days than not for the past 6 months).

Cognitive state anxiety is an unpleasant arousal from threatening demands or dangers. This condition is associated with many different bodily and cognitive expressions that are mostly undesirable.

Treatment for anxiety using drugs (anxiolytic drugs) is standard. Additionally, it is also known that physical exercise is effective in relieving anxiety and stress [23]. Due to relatively recent research, having different treatment avenues is a positive aspect of decreasing the risks of these serious diseases. Fear and anxiety are usually regarded to be harmful because they lower a person's quality of life, but it is also beneficial because a certain degree of anxiety increases a person's creativity and drives a person to perform. "He who has overcome his fears will truly be free." (From Aristotle).

Anxiety and cognition: The role of working memory

The results of studies demonstrate that induced anxiety deferentially impacts verbal and spatial working memory (WM). Also, low and medium-load verbal WM is more susceptible to anxiety-related disruption relative to high-load WM, while spatial WM is disrupted regardless of task difficulty.

Cognitive and defensive components of anxiety interfere with WM tasks but to a different degree. Anxious apprehension has more of a domain-general impact on overall WM. High-load verbal WM engages top-down control mechanisms that abolish anxiety-related disruption; spatial WM is more vulnerable to the effects of anxious arousal.

Cognitive state of anxiety

The model in Figure 1.3 depicts how exaggerated neurobiological sensitivity to threats in anxious individuals can lead to cognitive-behavioral threat responses characterized by a pattern of vigilance-avoidance, which ultimately results in sustained threat perception. Such sustained threat perception is accompanied by the prolonged activation of threat-related neural circuitry and threat-responsive biological systems, including the hypothalamic-pituitary-adrenal [15] axis, the autonomic nervous system (ANS), and inflammatory response. These system activations ultimately lead to elevated inflammation.

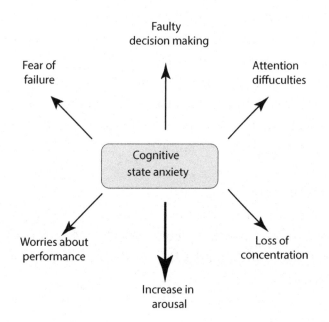

Figure 1.3 Undesirable effects of cognitive state anxiety. (Based on Wheelock, M., et al, 2014.) (Artwork by Liliana Cabrera).

Over time, the effects on central and peripheral systems may become chronic through structural changes in the central nervous system (CNS), altered sensitivity of receptors on immune cells, and accelerated cellular aging. Finally, such chronic elevations in inflammation can increase the risk for and accelerate the progression of the diseases associated with aging [16].

Disorders of fear and anxiety

Post-traumatic stress disorder

Additionally, fear may generalize the cues not associated with the traumatic events experienced by those people who go on to develop a fear-related disorder. In contrast, with resilience, fear responses to cues related to the traumatic event extinguish over time, and discrimination occurs between cues that are associated with the traumatic event and those that are not [24].

Postpartum depression and anxiety

Motherhood-specific concerns about the infant's vulnerability and safety are regarded to be completely normal, and even beneficial for infant wellbeing, but at high levels they negatively contribute to maternal anxiety [25]. Postpartum depression and anxiety are reported to occur in 10-20% of postpartum mothers. More is known about postpartum depression than postpartum anxiety, which often occurs together in the same person. Post traumatic anxiety does not have a unique diagnostic criterium which makes the reported rate of occurrence having large degrees of variability.

Panic disorders

Panic reactions are an outburst of signs of fear that may result from a severe and acute scare. These reactions are often in the absence of an ability to cope, such as the sensation of suffocation that can be experimentally induced by inhaling carbon dioxide (another experimental inducer of panic is an intravenous administration of lactate or cholecystokinin).

People with panic disorder show a dramatic down-regulation of $GABA_A$ receptors in the right insular cortex [26]. Similarly, social phobia patients show differential blood flow reduction in the insular cortex during a public speaking task. One study detected that the participants had a significant contiguous and global decrease in volume of the distribution of benzodiazepine-$GABA_A$ binding in the whole brain with peak decreases in the right orbitofrontal cortex, right insula, right lingual gyrus, left fusiform gyrus, right superior temporal gyrus, left middle temporal gyrus, right middle temporal gyrus, left dorsolateral frontal cortex, left anterior medial frontal cortex, and left frontal pole. (The participants in this study fulfilled the criteria for a *DSM-IV* diagnosis of panic disorder, and they had to have no parallel axis I or III diagnoses to be included.)

The study also showed that the participants had a global decreased benzodiazepine receptor binding which supported the hypothesis that panic disorder is due to defective brain inhibition that leads to or allows paroxysmal elevations in anxiety during panic attacks [27]. Individuals with a specific phobia showed an exaggerated right insular response to fearful faces (Wright et al. 2003). Patients with a generalized anxiety disorder (GAD) show reduced activation in the insular cortex after symptom reduction by Citalopram (Hoehn-Saric et al. 2004).

In another study comparing the BOLD-fMRI brain activity in 8 participants with social phobia along with of 6 control participants, the participants who were phobic had greater subcortical, limbic, and lateral paralimbic activity (pons, striatum, amygdala/uncus/anterior parahippocampus, insula, temporal pole) compared with the controls that were not phobic [27]. Specifically, the phobic participants had differential blood flow reduction in their insular cortex during a public speaking task. The phobic participant also had less cortical activity in regions important in cognitive processing, specifically the dorsal anterior cingulate and prefrontal cortex.

Role of deficits in social interactions

Some people have mental and developmental disorders that are associated with severe deficits in social interactions and communication. Disorders such as autism spectrum disorders (ASD) (autism and Asperger's syndrome) are often associated with social phobias; in fact, autism spectrum disorder has social anxiety as one of the primary symptoms. Anybody with neurodiversity of one kind or another, or people with any other difference or impairment may be prone to developing social anxiety [28].

Social Anxiety Disorders

Social Anxiety Disorder (SAD) is a form of phobia that is characterized by an overwhelming self-consciousness that manifests during everyday social situations. Social anxiety is caused by the worry of being judged by friends or peers or because of the risk of embarrassment in front of others [29]. The term "Social anxiety"' was coined by Janet (1903) to describe people who feared being observed while speaking, playing the piano, writing, or behaving in a way that would cause embarrassment in front of others. SAD is one of the most frequently diagnosed anxiety disorders [30].

The term social anxiety is now used to describe what many people experience in social interactions such as excessive fear, nervousness, and apprehension (Butler, 1999). [31].

A list of symptoms and signs that may be associated with social anxiety [28]:

1. Nervousness in meeting new people.
2. Anxiety in the presence of more than one person.
3. Performance anxiety.
4. Excessive self-criticism after episodes of anxiety.
5. Fear of upcoming meetings.
6. Negative self-talk.
7. Lack of self-compassion.
8. Self-consciousness.
9. Fear of being judged or evaluated by others.
10. Hypersensitivity to criticism.
11. Difficulty in speaking and expressing.
12. Stuttering and swallowing words.
13. Heaviness in the chest.
14. The sinking feeling in stomach, palpitations, sweating, tremors, nausea and more.

Clark and Wells model of social anxiety

One of the most common and well-supported models of social anxiety was described by Clark and Wells in 1995 and illustrated in Figure 1.4.

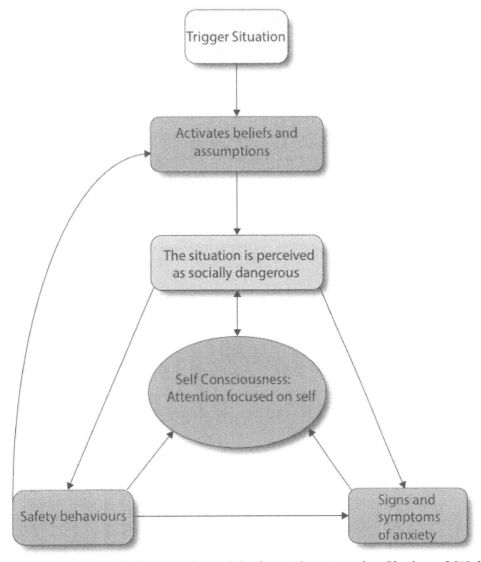

Figure 1.4 The cognitive-behavioral model of social anxiety by Clark and Wells 1995 (Based on Clark, D.M. and A. Wells, in *Social Phobia: diagnosis, assessment, and treatment*, R. Heimberg, et al., 1995 [28].) (Artwork by Liliana Cabrera).

When social anxiety elicits all the signs of activation of the sympathetic part of the autonomic nervous system, such as racing heart, sweating, clammy hands, trembling, stomach butterflies, and feeling sick, panic attacks and loss of consciousness may occur as a response to intense social anxiety. Social anxiety, in itself, can cause discomfort and embarrassment that often affects a person's ability to act naturally or perform a task in front of people. Stage fright or performance anxiety is a common aspect of social phobia which can trigger intense anxiety in connection with public appearance and performance.

Attacks of social anxiety may be accompanied by a person's mind going blank; a person may become confused and think that he/she has come across in a bad light. Many people who have such reactions to appearing in public and even perhaps in family appearances will often avoid social situations. This method of isolation results in poor self-esteem and depression.

Hypochondriasis

Hypochondriasis is a form of anxiety. It is the strong belief of having a severe disease or illness that can last a long time. It involves excessive worrying about having severe diseases without any signs or other objective indications of something being wrong with the person's health. Figure 1.5 shows a model of hypochondria.

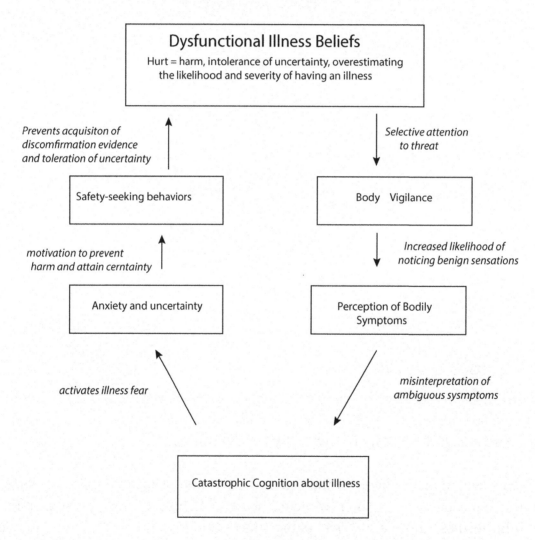

Figure 1.5 Model of hypochondria. (Artwork by Liliana Cabrera).

Munchausen syndrome

Munchausen syndrome is another complex mental illness where a person has severe emotional difficulties characterized by repeated false claims of having a physical or mental disorder without having any disease. The syndrome consists of feigning diseases, illnesses, or trauma to draw attention and sympathy. It often results in frequent hospitalizations and more extreme acts such as surgical operations and other treatments that carry high risks of severe complications. It is regarded to be a psychiatric disorder. There is a form of the Munchausen syndrome that may be regarded as Munchausen by proxy. It consists of abuse of another person in order to seek attention.

Angst

Angst is yet another feeling related to anxiety. The Danish philosopher Søren Kierkegaard introduced a translation of the word angst (from German) into Danish (angest) and from there into the English language. Angst describes an intense feeling of apprehension, anxiety, or inner turmoil which is often unfocused. The famous painting "The Scream" by the famous Norwegian artist Edvard Munch tried to represent "an infinite scream passing through nature". The Scream (1893) is often used to illustrate the conditions of angst.

The word angst (from German) is used in the English language to describe an intense feeling of apprehension, fear and anxiety, or inner turmoil. It seems that the word was first used outside the German language by the Danish philosopher Søren Kierkegaard (1813–1855). In "The Concept of Anxiety" (also known as The Concept of Dread, depending on the translation), Kierkegaard used the word Angest (in common Danish, angst, meaning dread or anxiety) to describe a profound and deep-seated condition.

"The Concept of Angst (Danish: Angest): A Simple Psychologically Orienting Deliberation on the Dogmatic Issue of Hereditary Sin, is a philosophical work written by Danish philosopher Søren Kierkegaard in 1844. The original 1944 English translation by Walter Lowrie (now out of print), had the title "The Concept of Dread". (The book is available on Questia: "The Concept of Dread," by Walter Lowrie).

"Should there then not remain uncertainty in fear and trembling until the last, I being what I am, and thou what thou art, I on earth, thou in heaven-a difference infinitely great-I a sinner, thou the Holy One? Should there not, ought there not, must there not, be fear and trembling till the last? Was it not the fault of the foolish virgins that they became sure, and went to sleep; while the wise virgins kept awake? However, what is it to keep awake? It is uncertainty in fear and trembling. Moreover, what is faith but an empty fantasy, if it is not awake? Moreover, when faith is not awake, what is it but that same precious feeling of security, which ruined the foolish virgins?" (Christian Discourses, Lowrie 1939 p. 219, Meditations from Kierkegaard, Translated and Edited by T.H. Croxall, The Westminster Press, copyright 1955, by W. L. Jenkins p. 56–57.)

Chapter 2 Emotions

Introduction

There are many ways to approach the understanding of fear and other emotions. One could focus on the psychological aspects of fear, or one could focus on the neuroscience of fear and what parts of the brain are involved in the expression of fear. Alternatively, one could focus on the fundamental changes that can be observed in an animal such as a rat when it is being scared, or one could focus on what happens in the human brain when a person experiences fear of one kind or another. Emotions and fear are fascinating topics to study because they have so many different sides; they can be beneficial, or they can be harmful to a person or to a society of people. In fact, fear and other emotions affect most people's lives a majority of the time. These different aspects of fear are discussed later in this chapter and subsequent chapters.

This chapter will go over general aspects of human emotion such as human value. This chapter will also discuss the components of a good quality of life and the prerequisite for success in life.

Fear is one of the six common emotions, anger, disgust, fear, happiness, sadness, and surprise. These are considered the primary or basic emotions and also referred to as archetypical emotions. Fear has many forms and many degrees of severity. It primarily serves to protect the body. The feeling of an immediate danger to the body, such as from a hostile person or animal, is one form of acute fear.

There are some doubts and disagreements regarding the exact meaning of the word emotion because different professional groups use the word differently. Throughout time, different authors who have written about fear have used different definitions of emotion. Also, different lexicons have different definitions for emotion. Merriam-Webster's dictionary has the word emotion defined as a strong feeling (such as love, anger, joy, hate, or fear). Another dictionary, Dictionary.com, states that the word emotion means an affective state of consciousness in which joy, sorrow, fear, hate, or the like, is experienced, as distinguished from cognitive and volitional states of consciousness (Figure 2.1 and 2.2).

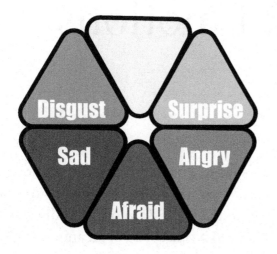

Figure 2.1 The primary or basic emotions and also referred to as archetypical emotions. (Artwork by Liliana Cabrera).

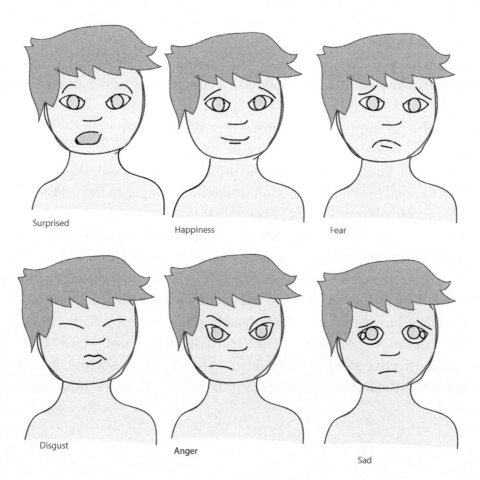

Figure 2.2 The primary or basic emotions and also referred to as archetypical emotions. (Artwork by Liliana Cabrera).

Emotions can have many different forms. The expression of some emotions, in the form of strong subjective feelings with specific facial expressions, are accompanied by the activation of the visceral motor system (visceral, from vicious meaning interior organ especially those in the abdomen). The body language connected to emotions is strong, and various specific facial expressions have been studied extensively in regard to their response to emotions such as joy, anger, and happiness.

Are emotions a psychological entity or is there a solid neuroscience basis for emotions? Barrett has discussed that question in a recent article [32]. She concluded that it is evident that emotions are real, but the question is, what kind of reality they are? In her article, she outlined a theoretical approach where emotions are a part of social reality, and she proposes that physical changes (in the face, voice, and body, or neural circuits for behavioral adaptations like freezing, fleeing, or fighting) transform into emotions when those changes take on psychological functions that cannot be perform by their physical nature alone. Furthermore, she states that this approach requires socially shared conceptual knowledge that the viewer uses to create meaning from these physical changes (as well as the circuitry that supports this meaning-making). She claims in her article emotions are, at the same time, socially constructed and biologically evident. Only when we understand all the elements that construct emotional episodes in social, psychological, and biological terms will we understand the nature of emotion.

Figure 2.3 Emotions are not (always) logical. (Artwork by Heather Trinth).

Theories of emotions

Many theories about emotions have been presented. The oldest, and perhaps the best known is the James-Lange Theory. This theory was presented by the American psychologist William James (1884) and the Danish physiologist Carl Lange (1887) who, at approximately the same time, independently proposed almost identical theories of emotion.

In short, the James-Lange Theory states that emotions are created by an event that causes arousal and, after interpretation, the emotion is created as a result. Other investigators have presented variations of this fundamental theory (Figure 2.4).

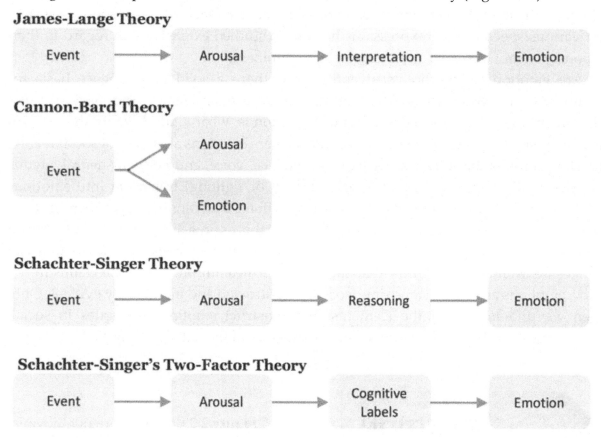

Figure 2.4 The oldest theory of emotions (top) and three other theories of emotion (Artwork by Heather Trinth).

The Cannon-Bard Theory assumes that the event that causes the expression of emotion has two parallel actions: Arousal and the creation of the feeling of emotion. The two other theories also assume sequential processing of the event that causes an emotional feeling. One theory, by Schachter-Singer, includes reasoning; the other, Two-Factor Theory, assumes that Cognitive Labels precede emotions. The main difference between these theories is the processing that occurs between arousal and the awareness of emotion. Only one of the theories, the Cannon-Bard theory, omit that step.

Figure 2.5 shows examples of how the view of a snake is processed according to the different theories of emotions that were shown in Figure 2.4. The end product is fear, and that is reached by different forms of intermediate expressions such as activation of the sympathetic nervous system (increased heart rate) or after an appraisal (Lazarus). Only the Cannon-Bard theory omits the intermediate step, and the emotion of fear is reached directly from the view of the snake.

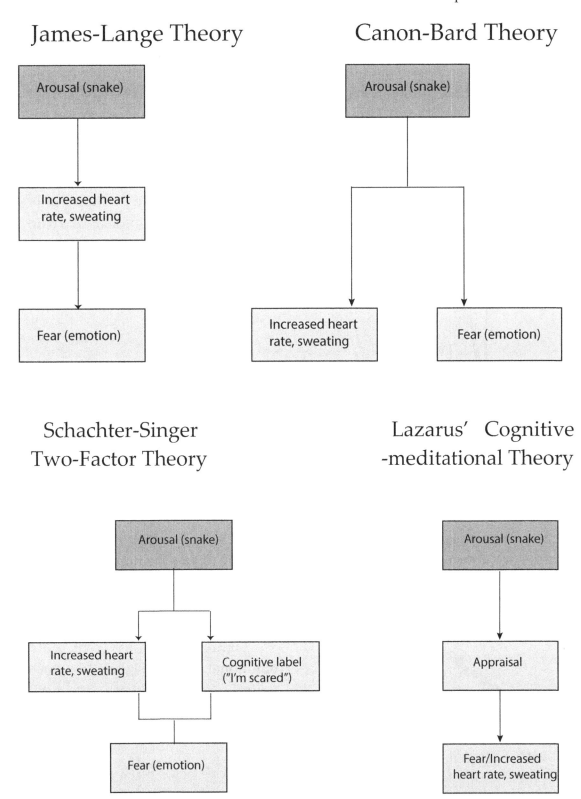

Figure 2.5 Comparing the theories of emotion. This figure illustrates how Lazarus' appraisal theory differentiates from the James–Lange, Cannon-Bard, and Schachter–Singer theories of emotion (Artwork by Liliana Cabrera).

Signals from the environment can trigger emotions that can affect many different bodily functions that are independent of from where the emotions are triggered (Figure 2.6). The physical reactions to emotions are mainly in the form of the activation of the sympathetic part of the autonomous nervous system. These reactions may be regarded as a direct expression of the various kinds of emotions. These bodily reactions can have many adverse effects and can even increase the risk of diseases.

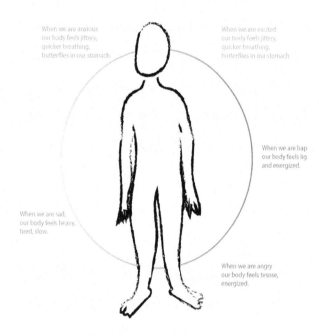

Figure 2.6 Some examples of bodily expressions of emotions. (Artwork by Liliana Cabrera).

As was mentioned above, information about events of various kinds can become controlling emotional responses directly or after appraisal (analysis). This is an example of how sensory information is traveling the short route (using the non-classical sensory pathways) or the long route (using the classical sensory pathways) (Figure 2.7).

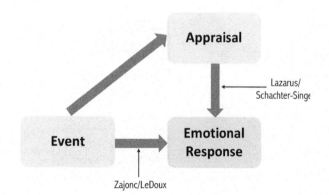

Figure 2.7 Two different routes for emotional neural signals to reach the neural circuits that generate an emotional response. (Based on Zajonc and LeDoux, 1984). (Artwork by Heather Trinth.)

According to Lazarus' cognitive-mediational theory, upon encountering a stressor, a person judges a potential threat (via primary appraisal) and then determines if useful options are available to manage and mitigate the situation (via secondary appraisal). Stress is likely to result if a stressor is perceived as threatening and few or no effective coping options are available [33] (Figure 2.8).

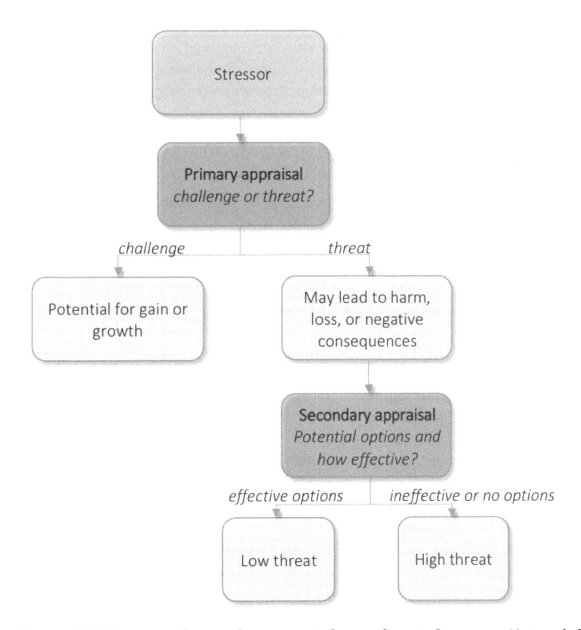

Figure 2.8 Primary and secondary appraisal according to Lazarus. (Artwork by Heather Trinth.)

Coping is a learned skill that is an essential way to manage diseases such as anxiety and fear. It embraces learning to live with a disease or condition such as anxiety, by understanding the condition.

Different definitions of emotion

A psychologist may define emotion as a complex state of feeling that results in physical and psychological changes that influence thought and behavior. Emotions such as fear come into a person's awareness, but is the experience and the awareness of an emotion, such as fear, a sensory experience? Some investigators think of and also include stress and anger as such emotions [21]. Does that mean that the experience and the awareness of emotion are similar to sensory experiences such as that from sounds, light, and odors or taste that reach a person's awareness? What about love and hate? What about hunger, thirst, and the illness experience, they also reach a person's awareness but are not regarded to be a sensory quality.

Some recent studies in humans have indicated that there are distinct differences between fear and other emotions, especially regarding visual expressions of emotions such as facial expressions. Among neuroscientists, the importance of understanding the neuroscience of emotions has led to a widespread interest in the neuroscience of emotions, and fear has become an increasingly important theme in modern neuroscience.

Love and hate

The love-hate relationship may be regarded as a kind of emotions. Hate is a passionate and robust form of aversion or dislike. It can have something to do with revenge or wish to revenge. People can hate many things, and what one person finds hateful differs from what other people find hateful. Some people dislike, and even hate, other people because they are different from themselves in terms of their different skin color, different religion, different origin (foreigners, xenophobia). In addition, many people dislike people with different sexual preferences such as gay, lesbian and other preferences (LGBT). This may be similar to the hostility between animals where, for example, laboratory rats that have wounds such as from surgical operations will get killed if they are in the same cage as normal rats.

Hate can have many causes

Zeki, 2008 [34] found that viewing a hated face resulted in increased activity in the medial frontal gyrus, right putamen, bilaterally in the premotor cortex, in the frontal pole and bilaterally in the medial insula. These investigators concluded that there is a unique pattern of activity in the brain in the context of hate.

Hate may be related to fear. Moreover, a person may hate what can imply danger to the person. Hate could be regarded as different forms of phobia. This kind of hate or dislike may have an equivalent in animals. Rats, for example, often kill animals in their own species if they are different in one way or another. It is unusual, otherwise, for an animal to kill members of its own species, thus very different from humans.

Human values

Core values

A person's core values are essential for having a satisfying life. A person's core values are those qualities or principles that are most valuable or desirable to a person. The ideals that a person believes in are important in the way a person lives and works. A person's values determine a person's priorities, and, deep down, they are probably a measure of a person's character. Here are some personal core values: Personal integrity, honesty, wisdom, respect for other people and their views, and a willingness to do something for others (without expecting a reward, altruism) are examples of personal core values.

Money is not one of the personal core values. A person's pursuit of money may be secondary to a person's core values, or a person may put money ahead of his/her core values. To what degree does money prevent a person to enjoy what is truly essential and meaningful to the person? Money can be a way to get a more comfortable and more secure life, but it may not make a person happier. Money may also mean power and influence. The perceived value of money seems to have the course of an inverted letter U (inverted parabola graph). The perceived value of money increases up to a certain dollar value and then it decreases. More money makes life easier in some ways, but the problems associated with being wealthy can overshadow the pleasure first associated with being economically independent.

Quality of life

There are many forms of quality of life, and different people perceive quality differently. Here is a list of what many people would agree upon as essential aspects of quality of life:

1. Freedom from diseases
2. Security: economic and personal
3. Satisfying work
4. Satisfying personal life
5. Do something for other people (Altruism)

Success in life depends on many factors that are different for different people, and it is often viewed differently by people of different ages. Here is a list of factors that many people would agree upon as essential for success in life:

1. Good genes
2. Luck
3. Wisdom and good judgment
4. Discipline, perseverance and the willingness to take risks.
5. Never be untruthful
6. Imagination and ideas.
7. Intelligence and a good education, marketable skills
8. Know how to act when things turn unfavorably

Different people have different goals in life, but most people want their goals to be fulfilled. Common goals may include:

1. Security: economically and personally
2. Obtain and hold an interesting and rewarding occupation
3. Be a good family person
4. Doing something for other people, for the community or the country, or the world?
5. Become Wealthy

How to fulfill a person's goals varies and many factors are involved such as what the interest of a person is in, what areas does a person have marketable skills, is the person prepared for the work, and what sacrifices that are necessary to achieve the goals in question a person may have.

Other personal qualities are **Happiness** - joy, contentment, pleasure, bliss, delight, and gladness. **Peace** - harmony, unity, tranquility, or serenity, **Perseverance** - persistence, or determination, **Respect** - appreciate, esteem, value, or cherish.

What people regret after a long life may provide suggestions about what a young person should consider.

1. "I wish I hadn't worried so much."
2. "I wish I'd had the courage to express my feelings more."
3. "I wish I had let myself be happier and more loving."
4. "I wish I hadn't held on to so much bitterness and resentment."
5. "I wish I hadn't been so afraid to follow my dreams."

(Gary Small MD, Psychiatrist, author of "Snap").

Chapter 3 Neurobiology of emotions

Introduction

The importance of understanding the neuroscience of emotions has led to a widespread interest in the neuroscience behind emotions, and fear has become an increasingly important theme in modern neuroscience. Neuroscientists usually define emotions as behavioral responses to environmental or internal situations. According to Lazarus' cognitive-mediational theory, a person will judge the potential threat when encountering a stressor, via primary appraisal, and then determine if effective options are available to manage the situation, via secondary appraisal. Stress is likely to result if a stressor is perceived as threatening and few or no effective coping options are available [33]. Additionally, activity in a threat-related neural network is elevated in people with anxiety disorders, as well as for persons exhibiting traits of high levels of anxiety [16]. How fear and other emotions are processed in the brain has been studied extensively by many investigators. Dr. Joseph LeDoux has been a pioneer in exploring the brain's processing of fear, anxiety and other emotions.

Fear activates many parts of the nervous system. The primary structures in the brain that are involved in creating an experience of fear are the amygdala, the hippocampus, and the prefrontal cortex. These structures together are known as the emotional brain (a term coined by Dr. Joseph LeDoux). Dr. LeDoux and others have confirmed the results of the old studies showing that structures in the brain known as the limbic system, especially the amygdala, anterior cingulate cortex, and prefrontal cortices, are involved in emotions. The amygdala activates many parts of the brain, which in turn activates many parts of the body. The activity of the central nucleus of the amygdala can activate the cerebral cortices, sharpen the senses, and facilitate retrieval of long-term memories that may be relevant to the emotion in question.

Since fear is an emotion, this chapter has a short discussion of the basic anatomy and physiology of emotions. Emotions, fear included, can be elicited by sensory stimulation. Sensory stimulation is information received from the outside or exteroception. Sensory systems are, therefore, involved in many forms of fear. Emotions can also be elicited by processes in the body, especially the brain or interception. Emotions can be evoked by "thinking" [35], or it can be related to memory.

The topic of affective, also known as mood-related, neurobiology is concerned with the neural bases of emotions and mood. The past 30 years have witnessed an explosion of research in affective neuroscience that has addressed questions such as:

1. Which brain systems underlie emotions?

2. How do differences in these systems relate to differences in the emotional experience of a person?

3. Do different regions of the brain underlie different emotions, or are all emotions a function of the same underlying brain circuitry?

4. How does emotion processing in the brain relate to bodily changes associated with emotion?

5. How does emotion processing in the brain interact with cognition, motor behavior, language, and motivation?

The neurobiology of human emotional processing spans cortical and subcortical structures. Experimental data regarding the organization of the neural networks involved in the processing of emotions, including fear, come from many different sources, including anatomical studies and physiological studies such as electroencephalography (EEG), magnetoencephalography (MEG), positron emission tomography (PET), functional magnetic resonance imaging (fMRI), and diffusion tensor imaging (DTI) [36, 37]. Studies of persons with diseases have also contributed to understanding the neuroscience of emotions. Studies of functional connectivity, a relatively new topic of neuroscience, has contributed much to our understanding of the underlying neurobiological processes of emotions including fear [38].

Neuroscience definitions of emotions

Neuroscientists define emotions as behavioral responses to environmental situations. As seen in Figure 3.1, the main components that generate the expression of emotions (learning and memory, fear and anxiety, and social behavior and stress) partly overlap because they receive input from the environment and send signals back to the environment. It is also indicated that these components are controlled by genetic, physiological, and neural mechanisms. Additionally, their functions can be altered by the activation of neuroplasticity.

Neural processes of emotion

The neural processes of emotions include genetics. For example, some people have stronger expressions of emotions than others, and activating emotions are easier for some people than for others. The neural activity in emotions, learning, memory, social behavior, and stress may interact in various ways to different kinds of emotional stimuli or situations resulting in the creation of a person's reaction (Figure 3.1).

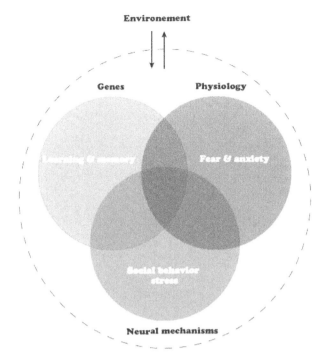

Figure 3.1 Neurobiology of fear, social behavior, learning, and memory. (Based on Jasnow Lab, Behavioral Neuroscience, 2015,) (Artwork by Liliana Cabrera).

To reiterate, the emotional neuro-pathway is initiated by an external (exteroception) or an internal (Interoception) stimulus that travels from the brain to the body, where physical changes take place, and back to the cerebral cortex, where information about the emotional response to the particular situation is coded and kept as a "feeling" (Figure 3.2).

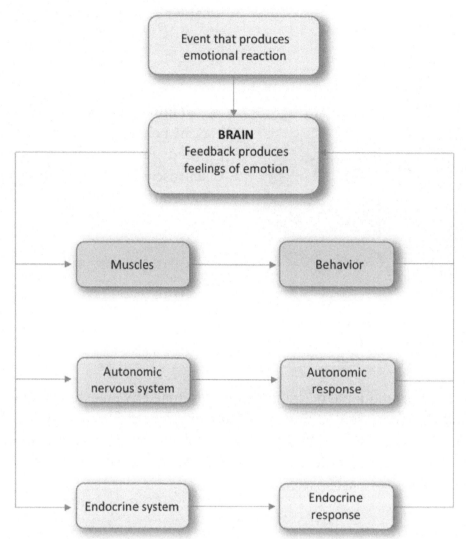

Figure 3.2 Interactions that may occur after an event that elicit an emotional response. (Based on: Isabel Jaén "The Romantic Syndrome: A Neuropsychological Perspective". Web site maintained by Isabel Jaen-Portello and Julien Simon [39]. (Artwork by Heather Trinth).

The role of threat on emotions

Threat perception leads to the activation of the hypothalamic-pituitary-adrenal axis, [15] which leads to an increased release of the glucocorticoid hormone, cortisol, from the adrenal glands. Threat perception activates the sympathetic arm, and also deactivates the parasympathetic division of the autonomic nervous system, leading to the increased release of catecholamines, epinephrine and norepinephrine. This pattern of activation and deactivation is accompanied by the increased synthesis and release of pro-inflammatory cytokines, including interleukin-1β (IL-1β), interleukin-6 (IL-6), and tumor necrosis factor-α (TNF-α).

The binding of these factors to receptors on immune cells regulates gene expression, including the expression of genes for pro-inflammatory cytokines. Thus, the effects of the hypothalamic-pituitary-adrenal axis and the autonomic nervous system on the immune system depend on the expression of immune cell receptors for cortisol and catecholamines, as well as on the release of these hormones. The glucocorticoid receptor appears to be down-regulated in response to a threat. This down-regulation limits the anti-inflammatory effects of cortisol. Although there are complex bidirectional relationships between the various factors in this model, threat perception ultimately leads to elevated inflammation levels [16].

Neural circuits involved in emotions

Three main systems are involved in emotions:

1. The amygdala, which receives inputs from structures in the brain including the association cortex of the temporal lobe.

2. The frontal cortex, the limbic system, and the olfactory system; the orbitofrontal cortex, whose inputs come from the other regions of the frontal lobes, temporal pole, amygdala, and limbic system.

3. The cingulate gyrus that projects to the limbic system and the frontal cortex [40].

Other parts of the brain that also play a role in creating emotional responses are:
1. Bed nucleus of stria terminalis
2. Frontal and prefrontal cortices
3. The thalamus (dorsal and medial parts)
4. Hippocampus
5. Hypothalamus
6. Periaqueductal gray

The general concepts of many brain functions have been recently revised, and it is now generally accepted that the compartmentalization of functions is less pronounced than earlier assumed. Instead of the idea of compartmentalization, the organization of the brain is more like a distributed system where there is a broad range of connections between different systems in the brain. Most connections are two-way, and that forms loops where information can circulate. Also, the concepts about the pathology of neurological disorders of various kinds have been revised recently. It is now assumed that network dysfunction plays an important role in the pathology of many neurological disordered [38].

This chapter covers some aspects of the neurobiology of emotions. How the parts of the brain that are involved in emotions can be reached by stimuli from the outside and the inside of the body is discussed. Many of the following descriptions of the neurobiology of emotions apply to fear. Proprioception, Nociception, Exteroception, Interoception--- What do they all mean?

Different kinds of signals can reach the emotional brain through many different routes. Exteroceptive signals are sensory signals such as visual images, sound, smell, taste, and touch coming from our surroundings. Interoceptive signals are those coming from inside the body, such as pain, hunger, thirst, and proprioception stemming from the muscles, joints, and more. These signals can reach the brain and elicit emotional responses of various kinds. The signals we receive from memory and "thinking" are also regarded as interoceptive signals.

Sources of information that may elicit emotions of various kinds

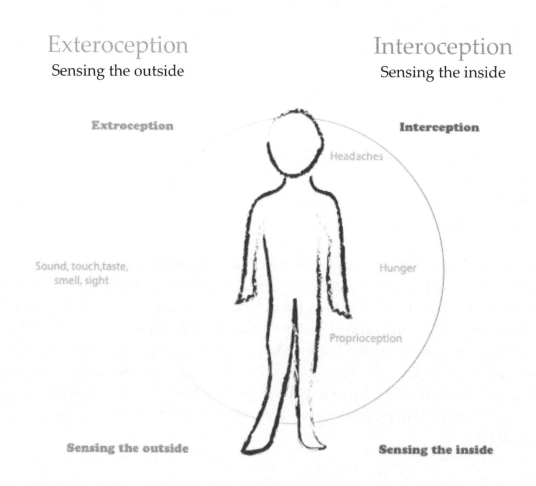

Exteroception
Sensing the outside

Interoception
Sensing the inside

Figure 3.3 Examples of sources of information that may elicit emotions of various kinds. (Artwork by Liliana Cabrera.)

The emotional brain

1. The discipline of affective (mood-related) neuroscience is concerned with the neural bases of emotion and mood.

2. LeDoux and his co-workers have discovered important functional pathways of fear, and this is now a hot research topic in neuroscience.

3. The parts of the brain that play the biggest role in emotions (including fear) was named *"The emotional brain"* by Joseph LeDoux, a scientist who has studied the neuroscience of emotions including fear extensively.

4. The emotional brain consists of several structures and it is a part of the limbic system.

Important brain structures that are active in emotions

The neuroscience of emotions is much more complex than earlier believed. It involves many different parts of the brain, many of which have earlier been assumed to have rather different and isolated functions. The connections between limbic regions and other structures are important for emotions, including fear and reward circuitries.

It is now generally accepted that the amygdala plays a central role in the organization of emotional responses and its translation into actions, respectively. Three main structures are involved in expression of emotions. The connections to and from the three main structures that are involved with emotions including fear and anxiety:

1. *The amygdala*, which receives inputs from the association cortex of the temporal lobe.

2. *The frontal cortex*, the limbic system, and the olfactory system; the orbitofrontal cortex, whose inputs come from the other regions of the frontal lobes, temporal pole, amygdala, and limbic system.

3. *The cingulate gyrus* that projects to the limbic system and the frontal cortex. The involvement of the cingulate cortex can explain some of the emotional components of pain, particularly centralized chronic neuropathic pain.

The cingulate gyrus is related to motivation, and other parts of the brain also play a role in creating emotional responses including the hippocampus, the different parts of the prefrontal cortex (PFC) and the hypothalamus. Several other structures play important roles in the creation of the experience of emotions and the effect induced by emotions such as fear. Recent studies also indicated that the amygdala might not be involved in all forms of fear. This is just another sign of the vast complexity of the central nervous system.

Three main systems are involved in emotion are located in the limbic system, the cerebral cortex. The sensory system plays an essential role in emotions.

Limbic system

The limbic system is a collective name used for a group of structures that are located deep in the brain near the edge of the medial wall of the cerebral hemisphere. The name limbic comes from the Latin limbus meaning rim. The different components of the limbic system play important roles in emotions including, fear and anxiety. Usually the hippocampus, amygdala, and fornicate gyrus are included in this term, but often structures to which they interconnect, such as the septal area, the hypothalamus, and a medial part of the mesencephalic tegmentum are regarded as belonging to limbic structures already. Also, limbic structures are sometimes referred to as the visceral brain.

The structures of the limbic system that are reached subcortically by input from the somatosensory system, specifically the lateral part of the of the amygdala system, are heavily involved in many functions including affective reactions to pain. The subcortical connection from the dorsal thalamus to the amygdala and other limbic structures may be responsible for these affective reactions to pain such as depression and fear. The projections from other structures to the insula can have complex consequences as well. The functions of the insula are poorly understood, but it may have intricate functions that are associated with the awareness of "self-sense" related to body ownership [41].

Connections from the cerebral cortex to limbic structures

The corticolimbic system has essential roles in emotionally related learning and the interactions between the basolateral amygdala, the dorsolateral prefrontal cortex, and the hippocampal formation. The corticolimbic system consists of several brain regions that include the rostral anterior cingulate cortex, hippocampal formation, and basolateral amygdala. The corticolimbic circuit integrates motivationally significant information such as pain and fear. The corticolimbic circuit also makes decisions about action selection. These structures are essential for pathologies especially schizophrenia [42].

Emotionally related learning is mediated through the interactions of the basolateral amygdala and hippocampal formation, and motivational responses are processed through the dorsolateral prefrontal cortex [42].

It has become evident that the fundamental transformation of the information that is represented in the lateral nucleus of the amygdala is sensory oriented and the code of information in the basolateral nucleus is behaviorally oriented. This transformation is assumed to be the gating role in regard to the input to the amygdala from the medial prefrontal cortex and associated cortical regions (further discussed in Chapter 4). The two areas of the mPFC, the prelimbic (PL) and the infralimbic (IL) areas, have a connection to the basolateral nucleus bidirectionally. These two areas have opposite effects. The PL is excitatory while the IL is inhibitory [43, 44].

Sensory systems

Sensory systems have a plethora of anatomical connections to the nuclei of the limbic system. Sensory input from the auditory, somatosensory, and visual systems can reach the emotional brain through two routes that lead to the lateral nucleus of the amygdala. One route, the "low route", is the subcortical connections from the dorsal and medial thalamic nuclei. The other route, the high route, consists of a long chain of neurons from the ventral part of the thalamus through primary, secondary, and association cortices that lead to the lateral nucleus of the amygdala.

Cells in the dorsomedial thalamus make many connections to limbic structures, particularly the hypothalamus and the lateral and basolateral amygdala nuclei. The cells in the medial thalamus that receive input from that part of the spinothalamic tract project to Brodmann's area 24c, which lie within the anterior cingulate sulcus.

Other parts of the brain also play a role in creating emotional responses such as the hypothalamus, which is an important output structure. Again, activity in a threat-related neural network is increased in people with anxiety disorders, as well as for persons exhibiting high levels of trait anxiety.

Dr. LeDoux has named these structures the "emotional brain." The term "emotional brain" according to the Oxford Dictionary originating in late 1900:

"The parts of the brain and mental processes involved in the experiencing of emotion; specifically, the amygdala and other parts of the limbic system."

It involves the structures of the limbic system, but the general opinion regarding which components of the limbic system that are involved in the emotional brain have changed, and different investigators have included different components of the limbic system in their understanding of the emotional brain. The first version of the limbic system was that of James Papez who published his concept of the emotional system in 1937. Joseph LeDoux published a revised version in 1994, and most recently Paré and Quirk (2017) published a more slightly modified version.

Before Dr. LeDoux's studies, James Papez (1937) described the limbic system and showed its involvement in emotions.

Papez's emotional brain
The Papez circuit theory (1937) of functional neuroanatomy of emotion

Papez' concepts of the emotional brain: Papez circle

James Papez, in 1937, described a part of the limbic system that he regarded to be involved with emotions. These structures, also known as Papez circle, form a rim around the corpus callosum, which is the large fiber tract that connects the left and right hemispheres of the brain. The structures that were first described as the limbic system are the cingulate gyrus and the hippocampus.

More recently, the circuitry described by Papez initially has been revised to include the orbital and medial prefrontal cortex, the ventral parts of the basal ganglia, the mediodorsal nucleus of the thalamus, and the three main nuclei of the amygdala including the bed nucleus [45].

Papez proposed that connections between the hypothalamus and the anterior thalamus and also from the cingulate cortex were the basis of emotions. He suggested that signals from the cingulate cortex reach the hippocampus followed by the hypothalamus allowing a top-down cortical control of emotional responses and then back to the cortex resulting in a trace of the emotional response to the particular situation being coded and stored as a "feeling."

Papez proposed that there were connections between the hypothalamus and anterior thalamus (Figure 3.4) and the cingulate cortex. Papez also suggested that the cingulate cortex integrates these signals from the hypothalamus, and that would be the basis for emotional experiences or feelings when the information reaches the sensory cortices. Papez argued that sensory messages concerning emotional stimuli that arrive at the thalamus are then directed to both the cortex, such as stream of thinking, and the hypothalamus, such as stream of feeling. Papez also suggested that the cingulate cortex integrates these signals from the hypothalamus, which would be the basis for emotional experiences or feelings when the information reaches the sensory cortices.

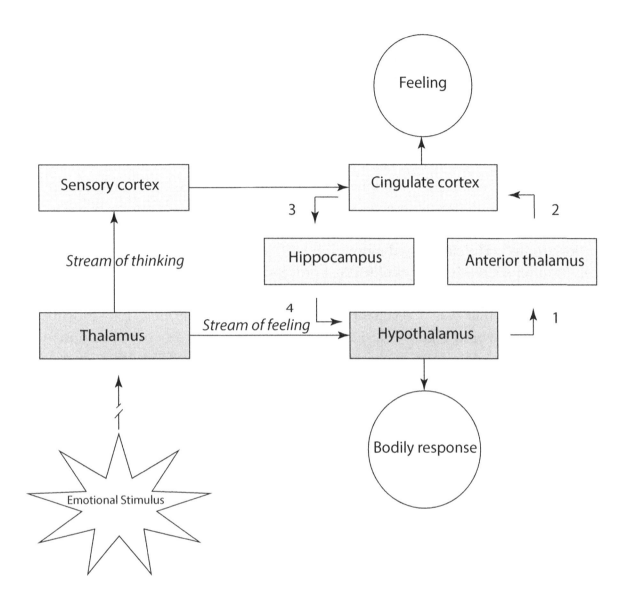

Figure 3.4 Papez's description of the circuit theory of the functional neuroanatomy of emotion. Notice that the amygdala is absent in this description of the emotional brain (Based on Dagleish, 2004.) [46]. (Artwork by Liliana Cabrera.)

Emotional experiences or feelings occur when the cingulate cortex integrates these signals from the hypothalamus with information from the sensory cortex. Output from the cingulate cortex (3) and then to the hypothalamus (4) allows top-down cortical control of emotional responses. According to Papez an emotional (sensory) stimulus may induce a feeling of emotion or provide a bodily response, such that of stress, by activating the HPA axis as shown [47].

LeDoux's description of the emotional brain

A modern conception of the emotional brain

Dr. LeDoux, his students, and his coworkers in the 1990s published a radical revision of understanding of the neural basis for emotions (LeDoux, 1998). LeDoux defined the amygdala as a center for emotional evaluation [48], specifically involved in fear conditioning. The thalamus receives rapid and crudely processed information from the dorsal-medial thalamus and the nucleus of the solitary tract (NST) which is referred to as the "low-route". The slower and more processed information from the ventral thalamus received, after successive stages of cortical processing, is the "high-route".

A description of the modern interpretation of the organization and function of the emotional brain that was a result of the work by Joseph LeDoux, his students, and coworkers is shown in Figure 3.5. LeDoux has called it the "emotional computer." To reiterate, the amygdala receives the quick and crudely processed information from the dorsal thalamus and the nucleus of the solitary tract (NST), and the amygdala also receives the more processed information from successive stages of cortical processing from the ventral thalamus. The amygdala plays a VITAL role in evaluating emotional stimuli.

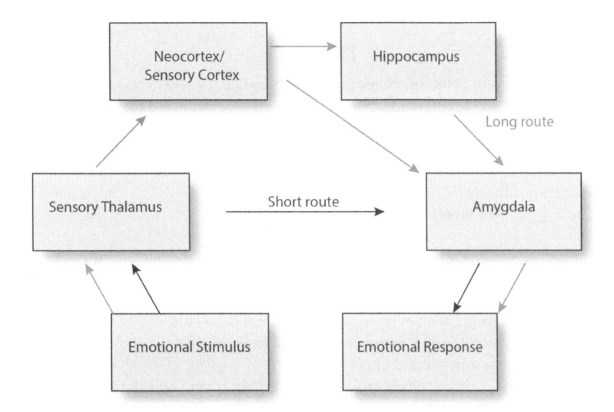

Figure 3.5 This hypothetical pathway shows the amygdala in the center of a network that is involved in emotions. Red arrows indicate the long route and black arrows show the short route. (Based on Dagleish, 2004.) [46]. (Artwork by Liliana Cabrera.)

Note that the amygdala can be reached through two different routes from the sensory thalamus, directly through a short route, the "low route", and through a long route involved a chain of cortical neurons, the "high route" [40].

Joseph LeDoux has published extensively regarding the neurobiology of emotions including fear. LeDoux has established that the amygdala is at the center of brain systems that are involved in emotions, and he has called it the "emotional computer" [49]. The model for the neural basis of fear, proposed by Joseph LeDoux, thus has the amygdala as the central structure, and the amygdala is now regarded to play a central role among the structures that are important for emotions. This is a major difference from the model proposed by Papez where the amygdala was not included in the description of the emotional brain network at all.

These differences between Papez's and LeDoux's descriptions can be summarized as:

1. The amygdala was not included in Papez's description.
2. The amygdala is the central structure in LeDoux's description.
3. Recent revision: The amygdala is only one of several structures that are activated by emotions (Paré, D. and G. Quirk) [44].

Revisions of the concept of the emotional brain

LeDoux's concept of the emotional brain was based on the research results that used the techniques of fear conditioning in animals. That research altered our concept of the emotional brain, from that of Papez's, by making the *amygdala nuclei the central structure of the emotional brain* together with the hippocampus, the periaqueductal gray (central gray), the lateral hypothalamus, the bed nucleus of the stria terminalis, and the deep superior colliculus.

The amygdala receives inputs from:

1. Sensory systems through the dorsomedial thalamus to the lateral nucleus of the amygdala.
2. Sensory systems through the primary, secondary and association cortex.
3. The frontal cortex, the limbic system, and the olfactory system;
4. The orbitofrontal cortex, whose inputs come from the other regions of the frontal lobes, temporal pole, amygdala, and limbic system.
5. The cingulate gyrus, which projects to the limbic system and the frontal cortex, and the bed nucleus of stria terminalis.
6. The frontal and prefrontal cortices.
7. The hippocampus.
8. The hypothalamus as an important output structure.
9. The periaqueductal gray.

Description based on fear conditioning

Specifically involved in fear conditioning, LeDoux defined the amygdala as a center for emotional evaluation To review, the amygdala receives rapid and crudely processed information from the "fast and dirty" pathway which stems from the thalamus and the nucleus of the solitary tract (NST), and the slower and more processed information is received from the "slow and accurate" pathway after successive stages of cortical processing (LeDoux, 1987; LeDoux, 1994).

The organization of the emotional brain, according to LeDoux, was based on studies using fear conditioning mainly in rodents (rats). A number of studies using these experimental methods were published over many years. Therefore, the neural networks involved are well characterized. These studies also represent cellular and molecular mechanisms which play a supporting role to the amygdala in the formation and subsequent consolidation of memory traces.

Some recent studies have shown that fear-based learning involves more complex circuits than initially proposed in the LeDoux concepts of the emotional brain. It was found that the amygdala is not the only structure where the activation of neuroplasticity is involved in what is known as Pavlovian fear conditioning [50].

Fear conditioning

In general, the technique of fear conditioning in animals has been the basis for much of our contemporary knowledge about the neuroscience of fear and anxiety. Fear conditioning is based on the work by the Russian scientist Ivan Pavlov. The Pavlovian fear conditioning is a form of learning where a neutral stimulus, such as a tone, is presented together with an aversive stimulus, such as an electrical shock to the foot of the animal.

The principle of fear conditioning is: In training, a sound is presented together with an electrical shock to the animal's foot (Figure 3.6). The neutral stimulus is the "conditional stimulus" (CS), the aversive stimulus is the "unconditional stimulus" (US), and the fear is the "conditional response" (CR). The movements of the rat are noted. After such training has been done for some time, the neutral stimulus alone will elicit the same reactions the animal had when the neutral stimulus was presented together with an aversive stimulus.

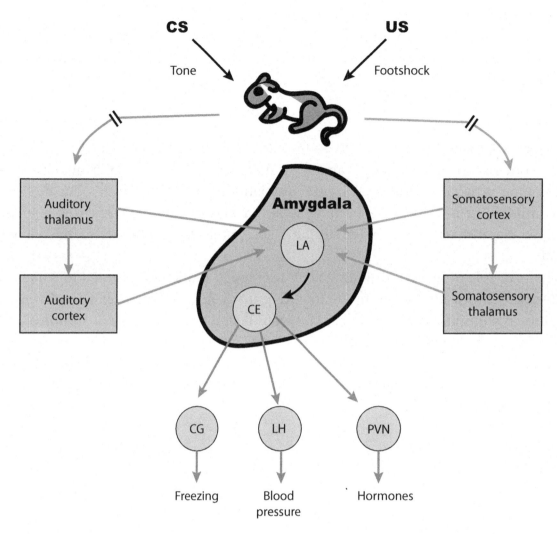

Figure 3.6 Method for fear conditioning in an animal. The reactions are shown in the form of a signal from the central nucleus of the amygdala. (Artwork by Liliana Cabrera.)

Recent revisions of the concept of the emotional brain

It was mentioned above that LeDoux's concept of the emotional brain was based on the results of research in animals on fear conditioning. The research by Joseph LeDoux and his colleagues changed our concept of the emotional brain from that of Papez making the amygdala nuclei the central structure surrounded by the thalamus, the sensory cerebral cortices, and the hippocampus. Also, the output from the central nucleus of amygdala to the periaqueductal gray (central gray), the lateral hypothalamus, the bed nucleus of the stria terminalis, and the deep superior colliculus has been noted.

It is also important that other brain areas such as auditory thalamic and cortical cells that projects to the amygdala exhibit increased responsiveness to conditioned stimuli [51]. Midline thalamic cells are also important in showing increased responsiveness [52]. Within the amygdala, several parallel inhibitory and excitatory circuits are involved. These are regulated by the medial prefrontal neurons activated during the expression of conditioned fear. The same circuits are activated during the extinction of conditioned fear [53].

Thus, recent studies show that that the amygdala is not the only location in the brain where the activation of neural plasticity for classical fear processing (Pavlovian) occurs, but also other structures are involved such as midline thalamic and cortical neurons [52].

The retrieval of fear memories is essential in human and animals. This has developed because it increases the chance of survival. Studies in rats have shown that the fear retrieval areas change over time [52]. These areas involve the dorsal midline thalamus (in rats) for late retrieval but not for early retrieval.

The prelimbic (PL) part of the prefrontal cortex, that is necessary for fear retrieval, sends large projections to the paraventricular nucleus of the thalamus which are involved in late retrieval but not in early. Additionally, eliminating the prelimbic input to the thalamus impaired early retrieval [52].

These studies thus showed that the paraventricular nucleus of thalamus plays an essential role in the consolidation of fear memory traces in the amygdala and in the retrieval of long-term fear memories. The study by Paré and Quirk (2017) [44], has confirmed the earlier assumption that the activity of neurons in the amygdala serve to signal threats and that these circuits generate the basis for defensive behaviors [44].

New description beyond fear conditioning

Classical fear conditioning is one of the most influential models for studying the neuronal substrates of associative learning and the mechanisms of memory formation in the mammalian brain [50]. It is now believed, however, that the models based entirely on the results of studies in rats on fear conditioning cannot wholly describe the neurology of fear.

Recent studies by Pare and Quirk (2017) [44], have shown that fear learning involves more complex circuits than initially proposed in the LeDoux concepts of the emotional brain [44]. These researchers showed that the amygdala is not the only structure where activation of neuroplasticity is involved in what is known as Pavlovian fear. Also, other structures are involved, such as midline thalamic and cortical neurons. The amygdala is involved in the evaluation of the significance of sensory stimuli and fear, and the amygdala rewards the use of similar neural structures. However, the cerebral cortex is not necessary for simple fear conditioning, but it allows a person to recognize an object by sight or sound.

On the other hand, pathways involving the cerebral cortex offer detailed and accurate representations of the environment. The old concept of the transformation that occurs between the lateral nucleus and the basolateral nucleus of the amygdala (Figure 3.7) was based on studies in rats using fear conditioning. This is in Figure 3.8 compared with a revised concept based on recordings from the basolateral nucleus [44, 54].

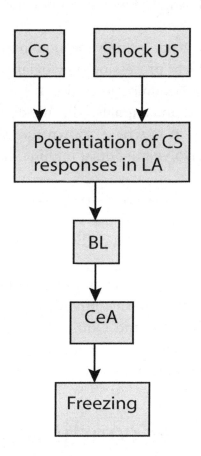

Figure 3.7 Old description of connections through the amygdala based on fear conditioning. Convergence of the tone (CS) and the shock (US) potentiates tone responses of neurons in the LA, which then activate BL amygdala neurons to drive freezing via the CeA. CS: Conditioned stimulation, US: Conditioned stimulation, LA: Lateral nucleus of the amygdala, BL: Basolateral nucleus of the amygdala, CeA: Central nucleus of the amygdala. (Based on Paré, and Quirk, 2017.) [44]. (Artwork by Liliana Cabrera).

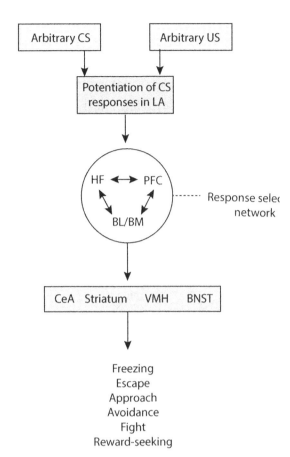

Figure 3.8 New description of connections through the amygdala beyond fear conditioning. Arbitrary CS-US pairings potentiate CS inputs to LA neurons, which project to a network of interconnected structures, including BL, basomedial nucleus (BM), medial prefrontal cortex (mPFC), and hippocampal formation (HF). This network selects an appropriate response from a variety of possible alternatives. Responses are executed through descending projections to CeA, striatum, ventromedial hypothalamus (VMH), or bed nucleus of the stria terminalis (BNST) (Based on Paré, and Quirk, 2017.) [44]. (Artwork by Liliana Cabrera).

The amygdala

In general, the amygdala processes sensory stimuli and its emotional value [55]. The amygdala is involved in many essential functions such as learned fear, aggression, social behaviors, and emotions [56]. The amygdala seems to be essential for the unconscious processing of fear based visual cues such as from facial expressions as concluded from studies in monkeys where the amygdala nuclei were ablated.

It was, however, not until in late 1970 that the results of more systematic studies of the function of the amygdala began to appear in scientific literature. Before that, little was known about the anatomy of the amygdala, and the function of the amygdala especially was not understood at all. With the advent of new techniques to study the function of the central nervous system, the number of studies of how fear is processed in the brain has increased rapidly.

The amygdala is a complex structure of anatomical and functional entities

The amygdala consists of three main nuclei. The main nuclei are the lateral nucleus, the basolateral nucleus, and the central nucleus, each with very different functions (Figure 3.9). The lateral nucleus of the amygdala projects to the basolateral nucleus, which connects to the central nucleus of the amygdala. The central nucleus is the main output nucleus of the amygdala, and the central nucleus neurons connect to the endocrine, behavioral, and autonomic centers of the brain. It is also the principal target of olfactory information.

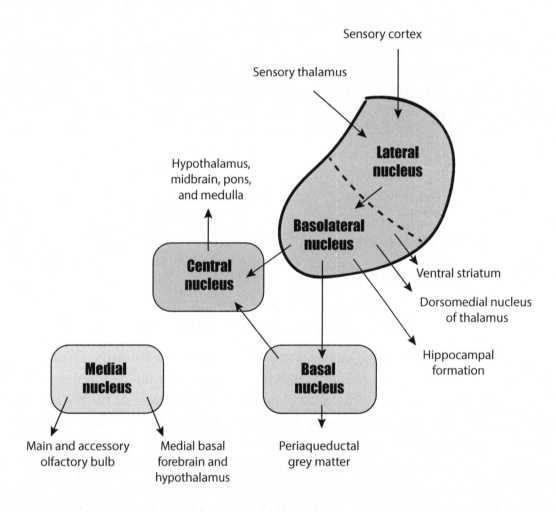

Figure 3.9 The main amygdala nuclei and their inputs and outputs are shown, emphasizing the complex internal architecture of this structure. (Based on: Adolphs, 2013.) [5, 57] (Artwork by Liliana Cabrera.)

Different authors have described the nuclei of the amygdala in different ways. The diagram in Figure 3.9 is an example of how one group of authors visualize the amygdala nuclei. These authors show the main components of the amygdala and their interconnections [57]. Here the basolateral nucleus and the lateral nucleus is shown as a unit, whereas the basal nucleus and the medial nucleus is a separate nucleus.

The bed nucleus of the stria terminalis often included in the term amygdala. The bed nucleus, the medial prefrontal cortex, and the infralimbic prefrontal cortex play important roles in creating the expressions and feelings of fear. Early studies in monkeys have shown that removing the amygdala in monkeys make them tame, thus a sign of reduced fear and aggression [58].

It is well-known that the basolateral nucleus (BLA) of the amygdala is the anatomical site for the integration of auditory, somatosensory, and nociceptive information in the amygdala. The same structures play a central role in the response of fear to discrete cues as well as its contexts. The BLA has vast reciprocal projections to the medial prefrontal cortex (mPFC). It has been suggested that mPFC regulates fear expression mainly through neural integration in the BLA. Contemporary developments and breakthroughs of techniques such as optogenetics and chemogenomic advances have made new studies of the involvement of these structures in fear possible.

A number of recent studies have suggested models of fear processing in the nuclei of the amygdala are considerably more complex than the older models [59]. It has been known for a long time that inhibition plays an important role in the normal and pathological functioning of the amygdala. The commonly used medications for the treatment of anxiety and fear are drugs such as the different forms of benzodiazepines discovered in 1955 by Stembach. These are GABA$_A$ receptor agonists. There is now a long series of molecules available that are all variations of one of the first synthesized benzodiazepines, diazepam (Valium).

Role of the extended amygdala

The "extended amygdala" includes the central and medial nuclei of the amygdala and the bed nucleus of the stria terminalis. These structures receive input from the basolateral amygdala and cortical regions such as the insular cortex. Memories related to fear are consolidated in the divisions of the prefrontal cortex, specifically the medial prefrontal cortex and the infralimbic prefrontal cortex.

The primary structures involved in cued and contextual conditioned fear expression and fear recovery are the amygdala, hippocampus, and the prefrontal cortex. People with phobias have a significantly increased amygdala activation than control groups under conditions of phasic fear. People with sustained fear had signs of increased neural activity in the insula and ACC, whereas people with a phobia had a stronger activation of the BNST and the right ACC under conditions of sustained fear than the control groups. People with a phobia have enhanced functional connectivity of the BNST and the amygdala.

The bed nucleus of stria terminalis

The bed nucleus of the stria terminalis coordinates autonomic and motor responses to threats, and the periaqueductal gray coordinates stereotyped defensive reactions to various threats, such as immobility and panic. In addition, evidence has been presented that show the activity in this threat-related neural network is potentiated for individuals with anxiety disorders, as well as for persons exhibiting high levels of trait anxiety ([16].)

Münsterkötter, A., et al., [60] studied phobias, specifically people with a phobia to spiders, and he showed that sustained anxiety has been linked with the activation of the bed nucleus of stria terminalis (BNST), anterior cingulate cortex (ACC), and the insula whereas increased amygdala activity has been associated with phasic fear [60].

People with phobias have significantly increased amygdala activation than control groups under conditions of phasic fear. People with sustained fear had signs of increased neural activity in the insula and ACC versus people with a phobia had a stronger activation of the BNST and the right ACC under conditions of sustained fear than control groups. Also, people with a phobia have an enhanced functional connectivity of both the BNST and the amygdala.

Münsterkötter's studies in humans have ultimately shown that the bed nucleus of the stria terminalis (BNST) is involved in some forms of fear and the effect increases when the distance between a spider and a person's foot is shortened [60, 61]. As a reminder, the gross neural circuits involved in cued and contextual conditioned fear expression and fear recovery are the primary structures involved are the amygdala, hippocampus, and the prefrontal cortex.

Routes to the amygdala

To reiterate, the amygdala can be activated from external sources (exteroceptive) and from within the body (interoceptive). The amygdala is a part of the limbic system, and the nuclei of the amygdala receive inputs from many structures of the brain, including the frontal cortex, other structures of the limbic system, and the olfactory system. The orbitofrontal cortex also provides input to the amygdala, whose inputs come from the other regions of the frontal lobes, temporal pole, amygdala, and limbic system. The orbitofrontal cortex also connects to the cingulate gyrus and then projects to the limbic system and the frontal cortex. The amygdala receives sensory input from two parts of the thalamus and the vagus nerve system, specifically the nucleus of the solitary tract (NST). Separate input comes from the hippocampus.

Information from thinking and stored information can reach the amygdala from the cerebral cortex and other high central nervous structures.

Sensory routes to the amygdala

Sensory systems play essential roles in emotions, and most signals that are associated with emotions are communicated in the form of touch, sound, vision (including facial expressions), or by odors which can also arouse emotions. Some studies in humans seem to indicate that visual signals are more important than other sensory signals for communicating fear. All sensory systems, except olfaction, have subcortical connections to the lateral nucleus of the amygdala from the dorsomedial thalamus, which belongs to the non-classical ascending sensory pathways [1]. Visual stimuli can also access the amygdala via a pathway that includes the superior colliculus and the pulvinar nucleus of the thalamus [62]. These routes are essential for eliciting emotions of various types and degrees.

External sources, such as hearing, somatosensory, and taste, are mainly in the form of sensory information that can reach the lateral nucleus, and for olfactory information relays to the central nucleus of the amygdala. Also, connections from cells in the dorsal-medial thalamus can reach the lateral nucleus of the amygdala directly through the association cortices, skipping the primary sensory cortices.

Studies using anterograde and retrograde tracing in primates and diffusion tractography in humans appear to indicate multiple (at least five) possible high-route pathways through multiple sensory cortices, as well as multi-area recurrence [36].

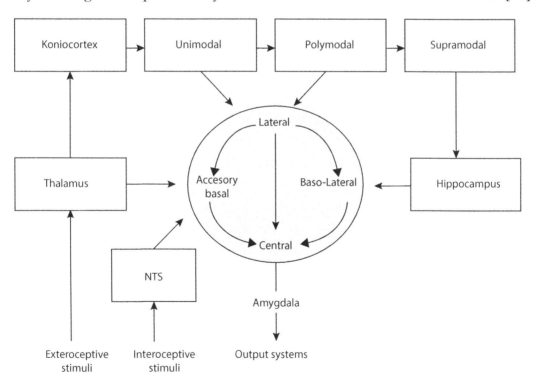

Figure 3.10   Some of the primary sources of input to the amygdala nuclei, emphasizing the complex internal architecture of this structure. (Based on Silverstein and Ingvar, 2015.) [36] (Artwork by Liliana Cabrera.)

Exteroceptive inputs reaching the thalamus through sensory systems are shown in Figure 3.10 to reach the amygdala through the vagus nerve and the nucleus of the solitary tract (NST). Information from the internal organs of the body also reaches the emotional brain through the vagus nerve and the nucleus of the solitary tract (NST).

Additionally, input from the hippocampus is also shown to reach the amygdala in Figure 3.10. Activation of neuroplasticity can affect the ability to activate the amygdala by sensory systems and from other structures.

Sensory routes to the amygdala

All sensory systems, except olfaction, have subcortical connections to the lateral nucleus of the amygdala from the dorsomedial thalamus, which belongs to the non-classical ascending sensory pathways [1]. Visual stimuli can also access the amygdala via a pathway that includes the superior colliculus and the pulvinar nucleus of the thalamus [62]. These routes are essential for eliciting emotions of various types and degrees.

Principally, sensory systems play a major role in emotion because most signals that are associated with emotions are communicated through most of the forms of our main senses such as sight, sound, touch, and smell. Olfaction has little cortical representation and the principal target of its pathway is the central nucleus of the amygdala. Some studies in humans seem to indicate that visual signals are more important than other sensory signals for communicating fear. The creation of fear from visual information has been studied in the brain of humans extensively [36].

Thus, all sensory systems project to the amygdala and it would, therefore, be possible for the amygdala to serve as the role of integrating information in all the different sensory systems. The amygdala may be especially necessary for the integration of taste and smell to create a holistic perception of the flavor of food.

Some studies have shown evidence that the non-classical auditory pathways may not be activated by sensory information in adult humans and, consequently, the subcortical route to the amygdala may not be functional under normal conditions in adults [63] [64]. However, there is evidence that the non-classical auditory pathways are active in children [64]. Findings have shown that some people with severe tinnitus have signs of activation of the non-classical auditory pathways which indicates that a subcortical route to the amygdala can become functional under pathological conditions through unmasking dormant synapses [63].

Two ascending sensory pathways

Sensory receptors are activated by a physical stimulus, followed by sensory nerves sending the information to brain's sensory nuclei, and then from there the two main parts of the thalamus are activated. Sensory systems, except olfaction, have two distinctly different ascending pathways in the brain. One is known as the classical pathway and one is known as the non-classical pathway. Joseph LeDoux called these two pathways the high route and the low route respectively. The classical route is also known as the slow and accurate route while the non-classical route is known as the "fast and dirty" route. Below are their anatomical pathways.

1. **High route:** Sensory nuclei > ventral thalamus > primary and secondary cerebral cortices > association cortices > lateral nucleus of the amygdala.
2. **Low route:** Sensory nuclei > dorsal and medial thalamus > secondary cortex > association cortices > lateral nucleus of the amygdala. (Figure 3.11, 3.12).

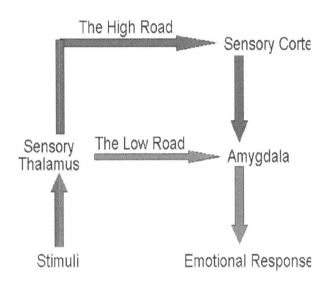

Figure 3.11 The two routes sensory stimuli may take to reach the amygdala.

Functional imaging studies have also provided evidence of increased activation in limbic structures in some people with tinnitus, which may occur through the "low route" [65]. There is also evidence from physiological studies that the non-classical auditory system may be abnormally active in select developmental disorders such as autism [66]. Certain sounds can cause affective reactions and that indicates the involvement of the amygdala. Whether that occurs through the classical or nonclassical pathways is unknown.

Fundamentally, two kinds of sensory information can reach the amygdala. The lateral nucleus of the amygdala receives extensively processed information through the ventral thalamus and a long chain of cortical circuits. The lateral nucleus of the amygdala also receives less processed or raw information directly through the dorsomedial thalamus.

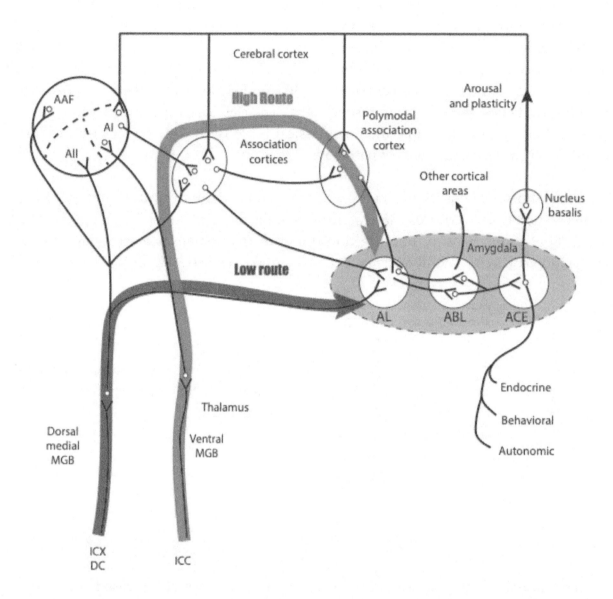

Figure 3.12 Two different routes to the amygdala from the auditory system showing both the "high route" with input from the central nucleus (ICC) of the inferior colliculus and the "low route" with input from the external inferior colliculus (ICX) and dorsal cortex (DC) of the inferior colliculus. Connections between the different nuclei of the amygdala and connections from these nuclei to different parts of the central nervous system are also shown. (Modified from Møller, 2014, based on LeDoux, 1992.) [67] (Artwork by Liliana Cabrera.)

In Figure 3.12, the "high route" receives input from the central nucleus of the inferior colliculus (ICC), and the "low route" receives input from the external nucleus (ICX) and the dorsal cortex (DC) of the inferior colliculus. The different connections between the nuclei of the amygdala to different parts of the central nervous system are also shown.

The two main pathways have specific differences; one is using the ventral thalamus and one using the dorsomedial thalamus as the pathway from receptors to the sensory systems, excluding the olfactory system. The high route offers detailed and accurate representations of the environment. The ventral thalamic pathway is slow and processes the information extensively and accurately. In this pathway, sensory signals reach the emotional brain through a long chain of nerve cells in different parts of the cerebral cortices where the actual analysis occurs. (For a detailed description of the sensory pathways, see [1].)

High route: Sensory nuclei > ventral thalamus > primary and secondary cerebral cortices > association cortices > lateral nucleus of the amygdala.

The low route is the less processed and raw information from sensory organs that travels in the non-classical sensory pathways consisting of the dorsal-medial thalamus and from there directly to the lateral nucleus of the amygdala and other subcortical structures. Information from the dorsomedial thalamus can also reach the lateral nucleus of the amygdala through a route consisting of the secondary cerebral cortex, bypassing primary cortices, connecting to association cortices and then to the lateral nucleus of the amygdala.

Low route: Sensory nuclei > dorsal and medial thalamus > secondary cortex > association cortices > lateral nucleus of the amygdala.

The dorsal and medial auditory thalamic nuclei that are part of the low route mainly receive input from the midbrain auditory nuclei (IC), but it has recently been shown that axons from cells in the dorsal cochlear nucleus (DCN) travel uninterrupted to the thalamus and make synaptic connections with cells in the medial geniculate body (MGB) [68]. This direct auditory route from the DCN to the MGB provides an even shorter and faster route for auditory information to reach the amygdala than the "low route" through the IC [6, 69].

A more detailed drawing of the two ascending auditory sensory routes to the amygdala is shown in Figure 3.13.

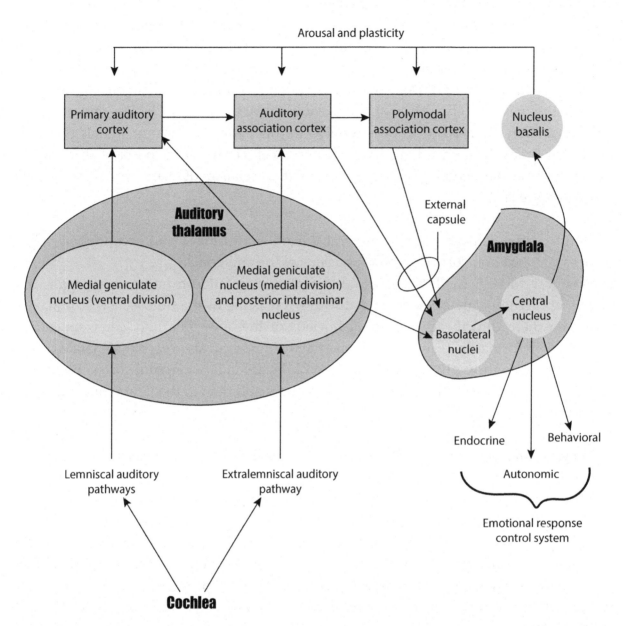

Figure 3.13 The two routes (high and low routes) that auditory stimuli can take to the amygdala is particularly important in emotional learning. Extensive output pathways from the central nucleus of the amygdala make cholinergic projections from the nucleus basalis to the cortex can arouse the cortex. (Artwork by Liliana Cabrera.)

A subcortical pathway to the amygdala

Most of the information that forms the basis of contemporary understanding of the sensory pathways to the emotional brain is based on studies in animals. There is, however, considerable evidence from many studies in humans that the amygdala can be activated through the subcortical pathway that goes through the dorsal-medial thalamus (the non-classical pathway or low route). For example, studies of people who had lesions in the primary visual cortex show that the amygdala can be activated without cortical processing [62]. (For details about the anatomical organization of sensory systems, see [1].)

The low route is the fastest, entirely subcortical, and it has been proposed to be unconscious [62], although some investigators have refuted that position [8]. The pathway of the high route includes primary sensory and association cortices. This pathway is slower than the low route, but it may provide us with a conscious perception of reality.

Evidence from studies of people who are blind indicate that the low route perceptual signals travel from the retina to the superior colliculus, back to the pulvinar thalamus, all before arriving at the lateral nucleus of the amygdala [70] [71] [3]. Anatomical studies using the tree shrew show projections from the superior colliculus to the lateral amygdala via the dorsal pulvinar of the thalamus. Using diffusion tensor imaging (DTI), one can map white matter in the brain, and it has been shown that there are connections between the superior colliculus and the amygdala via the pulvinar in human studies that are in vivo, [7].

Pathways from the dorsal-medial thalamus to the amygdala (the low route) provide only a crude perception of the world, but because they involve only one neural link, they are fast pathways. Fast pathways are essential because they provide quick reactions to potential danger and has been essential for the survival of many animal species.

Evidence of activation of the non-classical sensory pathways in humans

Abnormal access to the amygdala through the non-classical auditory pathways may occur in diseases in adults, such as severe tinnitus which has been associated with abnormal activation of non-classical pathways [72, 73]. This activation may result in the opening of subcortical connections to limbic structures, which may be the reason why some individuals with severe tinnitus experience the fear of sound or phonophobia [72]. Other studies have shown indications that the non-classical auditory pathways are active in young children but rarely active in adults [64].

The fast pathway may prepare the amygdala to receive more highly processed information from the cortex. As an example, when a slender spiral shaped object is located behind a tree in the visual field of an animal or human, it is much better to jump back and later recognize that it was a harmless object than to fail to quickly jump back in case it actually was a snake.

Auditory information, like the previously discussed cortical information, can also reach limbic structures through the primary auditory cortex and association cortices via the "high route" [67] [55]. Again, this route communicates highly processed information to the amygdala, compared with the "low route" that communicates faster, carrying less possessed information to limbic structures.

Recordings from specific nuclei of the amygdala in monkeys, mostly the lateral nucleus, found that the majority of the neurons, about 64%, responded to both the identity of the animal and its facial expressions, suggesting that these parameters are processed jointly in the amygdala [74]). A large proportion of the cells that were studied responded to the identity of the animal.

Responses of some cells to appeasing faces were marked by significant decreases in firing rates, whereas responses to threatening faces were strongly associated with an increased firing rate. These investigators interpreted their results to indicate that threatening faces may produce a more extensive activation in the amygdala, as opposed to the level of activation a neutral or appeasing face may produce.

Importance of visual information

Visual information about emotional facial expressions activates structures of the limbic systems, while information related to identity recognition is processed in the prefrontal cortex. While some mammals use facial expressions to show anger, hostility, and more, the human use of facial expressions is more complicated than it is in animals. We are capable of making and identifying many more forms of smiles and other various expressions of emotions than, for instance, cats and dogs. It is not known if that means that vision and hearing have stronger connections to limbic structures in humans than in many animal species that rely on olfaction for such communication.

Different forms of facial expressions seem to be processed by different populations of cells. Adolph and his colleagues have shown evidence that facial expressions related to fear are processed in different populations of cells than those that process information regarding the identity of faces [75]. This is a form of stream segregation. Unlike unilateral destruction, bilateral destruction of the amygdala removes the component of fear from the expression of faces but leaves the ability to recognize faces unimpaired. This indication means one stream carrying fear information uses the amygdala while the other stream carrying innocuous information about the identity of a person uses the visual nervous system [75, 76].

The high road (cortical) and low road (subcortical) for the visual system

On the high road, visual information from retinal ganglion cells is relayed to the visual cortex via the lateral geniculate nucleus, which is an area in the thalamus. Visual information is processed through several areas of the cortex before it is sent to the amygdala. After arriving in the amygdala, the autonomic and endocrine mediator of fear is engaged.

On the low road, visual information is sent first to the superior colliculus in the midbrain before being relayed to the amygdala via the pulvinar thalamus [77] (Figure 3.14). Unfortunately, little is known about the function of the pulvinar thalamus. The pulvinar thalamus is small in sub-primates, and it increased rapidly in size through the evolution of primates to reach a larger size in humans where it occupies as much as forty percent of the volume of the thalamus.

High road

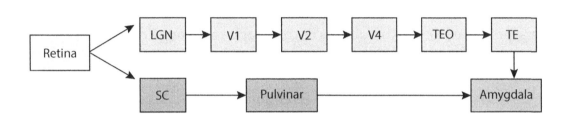

Low road

Figure 3.14 The high road and the low road from the retina in the eye. LGN, lateral geniculate nucleus; SC, superior colliculus; TE, inferior temporal cortex; TEO, inferior temporal cortex; V1, 2, 3, 4, visual cortex. (Based on Pessoa, Adolphs, and Rev, 2010 [8]. (Artwork by Liliana Cabrera).

There is evidence that the neuroendocrine system's response to visual threats has remained relatively unchanged for hundreds of millions of years [77], but it is not clear which route visual information takes to the emotional brain. How visual information is relayed to telencephalic areas in fish, that is homologous to the central nucleus of the amygdala (CeA) in humans, is unknown. Primates appear to depend less on subcortical visual pathways than other mammals, but studies continue to indicate a role for the superior colliculus and pulvinar thalamus in fear activation. However, it is not clear how these areas communicate with the emotional brain [8].

Morris et al. (1996) reported that that PET scans showed fearful visual stimuli evoked a response in the amygdala in human volunteers [78]. Changes in blood flow in the left amygdala, but not the right amygdala, were detected when fearful faces were presented in contrast to the presentation of happy faces.

Studies in monkeys, using recordings from cells in the amygdala and the ventral prefrontal cortex, showed cells in the amygdala were activated upon presentation of a scream facial expression, which is strongly negative. Opposingly, cells in the vlPFC were activated when a "Coo" was presented.

A "Coo" is a facial expression with multiple meanings depending on the social context [79]. These investigators interpreted their results to suggest that that the amygdala processes strong emotions roughly but rapidly; whereas processing ambiguous facial information in the vlPFC is slow and, therefore, provides the basis for an accurate decision. This is thus similar to the processing of information in the classical and nonclassical sensory pathways.

Routes to awareness of sensory information

Less processed or raw information can reach the lateral nucleus of the amygdala directly from the dorsomedial thalamus, where cells project directly to the lateral nucleus of the amygdala and many other subcortical structures. The cells in the dorsomedial thalamus also send connections to the cerebral cortex, but these connections bypass the primary cortices and connect to cells in the secondary and association cortices. These pathways are also known as the non-classical sensory pathways [1].

Both, processed information and raw sensory information, can reach the neural circuits that control awareness. To review, heavily processed information from sensory organs reaches the lateral nucleus of the amygdala and other subcortical structures through the classical ascending sensory pathways, via the "high route", from the ventral thalamus through a long chain of cells in the primary and secondary and association cortices. These pathways are also known as the classical sensory pathways.

Other ways sensory information can reach the amygdala

Axons from cells in the sensory nucleus of the vagus nerve (nucleus of the solitary tract, NST; also known as the nucleus tractus solitarious, NTS) can reach cells in the lateral nucleus of the amygdala. This pathway may integrate the olfactory experience with taste and visceral activity. Also, this pathway modulates visceral functions by emotional stimuli.

Recent findings have indicated that there are uninterrupted connections between cells in the dorsal cochlear nucleus and the medial part of the auditory thalamic nucleus (MGB) which provide an even shorter and faster route for auditory information to reach the lateral nucleus of the amygdala [68].

Sensory signals can also reach the amygdala from the olfactory bulb. The primary targets of axons in the olfactory pathways are cells in the central and the accessory basal nucleus of the amygdala [80]. Additionally, sensory signals can reach the amygdala from cells in the cerebral cortex terminating in the lateral nucleus and the accessory basal nucleus of the amygdala.

The accessory olfactory bulb that receives input from the vomeronasal organ projects exclusively to the cortico-medial nucleus of the amygdala.

Most of these pathways to the amygdala are reciprocal, indicating that the amygdala can modulate processes in other parts of the brain including those involved in the processing of sensory stimuli. Specifically, many cells in the amygdala project to the hypothalamus and the brainstem [80]. The author of these studies, Dr. Price, and the pictures use slightly different terminology for the three main divisions of the amygdala.

Connections from the amygdala nuclei to other structures

There are connections from the amygdala to many structures in the brain such as the hypothalamus, reticular formation, PAG, and the nucleus of the vagus nerve referred to as the nucleus tractus solitarius (NST). The connections from the amygdala are abundant and mostly reciprocal to the afferent connections (Figure 3.15). As is common for most connections in the brain, almost all connections to and from the amygdala nuclei are reciprocal indicating that the amygdala can also modulate sensory processing.

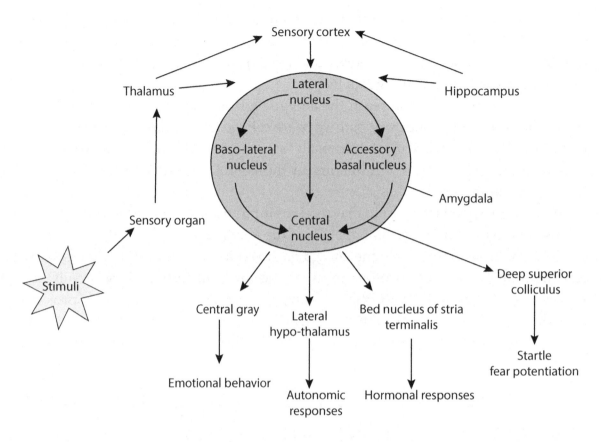

Figure 3.15 Connections to and from the amygdala emphasizing the connections to other brain structures. (Based on Biological Psychology, Fourth Edition, (Part 3), 2004). (Artwork by Liliana Cabrera).

The central nucleus connects to cortical regions such as the prefrontal cortex, the structures of the brainstem, and the hypothalamus [55]. Also, the hypothalamus and visceral sensory structures, such as the NST, are essential targets for these nuclei. The brainstem structures to which the central nucleus projects are involved in emotional, endocrine, and autonomic reactions. The connections to the paraventricular hypothalamic nucleus can cause an increase in the secretion of the adrenocorticotropic hormone, ACTH, along with increased secretion of corticosteroids, which are typical components of "stress responses."

As seen in Figure 3.15, there are connections from the amygdala to many structures in the brain such as the reticular formation, the periaqueductal gray (PAG) (central gray area), and the nucleus of the vagus nerve (NST). Specifically, many cells in the amygdala project to the hypothalamus and the nuclei in the brainstem [80]. Most of the pathways to the amygdala are reciprocal which indicates that the amygdala can also modulate processes in other parts of the brain, including those involved in the processing of sensory stimuli [80]. Note: A slightly different terminology is used in Figure 3.16 for the three main divisions of the amygdala than other pictures shown in this book.

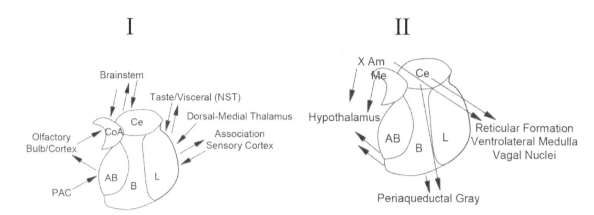

Figure 3.16 Comparison between the input to the amygdala and the output from the amygdala. I: Sensory input to the amygdala. II: Summary of outputs from the amygdala to the brainstem.

AB: Accessory basal amygdaloid nucleus; B: basal amygdaloid nucleus; Ce: central amygdaloid nucleus; CoA: anterior cortical amygdaloid nucleus; L: lateral amygdaloid nucleus; Me: medial amygdaloid nucleus. (Based on Price, J.L., 2003.) [80]. (Artwork by Monica Javidnia).

The amygdala is the brain site most critical for fear learning. Within the amygdala, the critical plasticity mechanisms underlying the acquisition of fear conditioning are thought to occur in the lateral amygdala and the lateral portion of the central nucleus (CEl). The medial division of the central nucleus of the amygdala (CEm) projects to various brain areas that produce fear and panic symptoms as seen in people with fear-related disorders.

The connections to and from the amygdala are many, summarized in Figure 3.17 [24]. The output of the amygdala nuclei to other parts of the brain are mainly mediated through the central nucleus (CE).

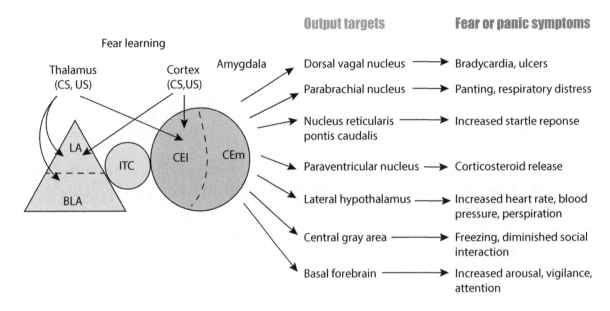

Figure 3.17 Schematic depicting the amygdala, the brain site most critical for fear learning (Based on Parsons and Ressler, 2013) [24]. (Artwork by Liliana Cabrera.)

The central nucleus also projects to the nucleus basalis, which is vital in arousal and promoting neural plasticity. Arousal and increased vigilance are, therefore, typical reactions of the activation of the central nucleus of the amygdala. Projections from the amygdala to brainstem nuclei can potentiate reflexes such as the startle response. These projections also mediate arousal through connections to the reticular formation. Another target is the lateral hypothalamus through which activity from the amygdala can cause a sympathetic activation response such as elevated blood pressure, tachycardia, pupil dilation, and more.

Connections from the basolateral and central nuclei of the amygdala that go to the nucleus of the solitary tract (NST) (nucleus of the afferent vagus nerve) cause a parasympathetic response activation that may result in bradycardia, increased intestinal activity, and more. Other projections from the nuclei of the amygdala include the periaqueductal gray (PAG). The stria terminalis, a band of fibers that cover the thalamus, has also been identified by using anatomical techniques [55, 81]. In addition, the connections to the PAG may mediate pain sensations related to fear. Structures belonging to the allocortex, especially the hippocampus, septal nuclei, nucleus accumbens, and the nucleus basalis, also receive input from the amygdala.

Only recently has the functional connectivity between different structures in the brain been studied with methods that allow the determination of the strength in the connections between different structures in the brain [82].

The brainstem structures, in which the central nucleus project to, are involved in emotional, endocrine, and autonomic reactions. To review, the connections to the paraventricular hypothalamic nucleus can cause increases in the secretion of the adrenocorticotropic hormone (ACTH) which, in turn, increases the secretion of corticosteroids, which are typical components of "stress responses." Any of these reactions may be reduced by injuries to the central nucleus of the amygdala. The central nucleus also projects to the nucleus basalis of the forebrain cholinergic system. This system is essential in arousal and for facilitating activation of neuroplasticity.

There are ample reciprocal connections between the nuclei of the amygdala and the medial and dorsal thalamic nuclei through which the output from the amygdala may reach prefrontal cortices. Recent studies of the auditory system have shown connections from the amygdala to the central nucleus of the inferior colliculus (ICC) [83]. Thus, the structure has been associated with classical sensory pathways, specifically the auditory pathway. The projection from the amygdala to the central nucleus of the IC (ICC) may close a loop consisting of the dorsal cortex (DC) and the external inferior colliculus (ICX), which connects to the medial and dorsal MGB of the thalamus, the basolateral nuclei via the lateral nucleus of the amygdala, and to the central nucleus of the amygdala.

Most of the connections to and from the amygdala are identified using anatomical methods, and the functions of only a few of these connections are known in detail. The synapses that establish the functional connections may be closed and dormant but have the possibility and ability to open under certain circumstances such as through the expression of neural plasticity [84].

Functional aspects of the amygdala

The amygdala plays a prominent role in the complex creation and regulation of emotions in general. The amygdala activates many parts of the brain, and then those different parts control the functions of many parts throughout the body. Signals from the central nucleus of the amygdala can activate the cerebral cortices, sharpen the senses, and facilitate the retrieval of long-term memories that may be relevant to the specific emotions in question [40].

The amygdala and the orbitofrontal cortex play essential roles in the organization of emotional responses and the translation into actions. The amygdala is now regarded to have a prominent duty for emotions in general [85, 86], and the amygdala's close extensions, known as the bed nucleus, play a central role in all anxiety and fear-like conditions in harmony with the connections from the central nucleus of the amygdala that connects to many structures in the brain, especially the prefrontal cortex [22].

The amygdala nuclei are anatomically close to the bed nucleus, which may be regarded as an extension to the amygdala. In addition, many other structures, especially the prefrontal cortex, play a central role in all anxiety and fear-like conditions. Similarly, the cingulate gyrus is related to motivation.

The amygdala, being a part of the limbic system, is a part of the old brain. It affects many systems of the brain, and the amygdala can also modulate many systems such as the sensory systems [87]. Specifically, the amygdala induces the arousal and the promotion of the activation of neuroplasticity.

The basolateral nucleus of the amygdala, through its projections to the central nucleus of the amygdala and the nucleus basalis of the forebrain, promotes the arousal of both primary and association cortices. The nucleus basalis (of Meynert) also facilitates neural plasticity, which may change the connectivity of the cerebral cortex [84, 88-91] [1].

Information processing in the amygdala

The amygdala receives both processed and less processed information from the sensory systems, and the amygdala nuclei perform further processing of the information as it receives the information. It then makes decisions such as regarding threats (Figure 3.18).

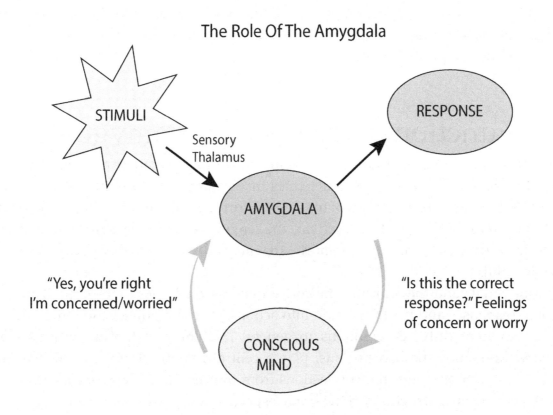

Figure 3.18 The amygdala, emphasizing its ability to process and evaluate stimuli. (Artwork by Liliana Cabrera.)

The amygdala nuclei communicate with the brain's networks that may be best described as the mind. The amygdala makes two-way connections with structures that have to do with consciousness, and the loops that are formed with these connections can cause processing of information of iterative nature.

The amygdala as a regulator of sensory functions

The amygdala nuclei have many different functions, and one of these is the regulation of sensory processing. For example, deep pressure has a direct influence on the amygdala. As a result, the amygdala has a regulatory effect on sensory processing. The amygdala reduces the overwhelming effect of all the stimuli received.

The amygdala also helps calm and soothe an over-stimulated or 'anxious' nervous system which then provides a safe and protected feeling. This can allow a person to better cope with stress and overstimulation, reduce meltdowns, improve concentration, and fall asleep quicker. This may occur in sensory systems as early as the receptor level.

Studies of the olfactory sensory system have shown that joint pharmacological inactivation of the central, basolateral, and lateral nuclei of the amygdala strongly suppressed odor-evoked activity in the GABAergic inhibitory interneuron populations in the olfactory bulb (OB) [87]. This suppression was prevented by the inactivation of the locus coeruleus (LC) in mice or the pretreatment of the olfactory bulb with a broad-spectrum noradrenergic receptor antagonist. Frost and McGann concluded that amygdalar activity influences olfactory processing as early as the primary sensory input to the brain. This is done by modulating norepinephrine release from the locus coeruleus into the olfactory bulb [87].

These authors interpreted these findings to indicate that the amygdala and the state of the LC actively determine which sensory signals are selected for processing in sensory brain regions. Since similar local circuitry that operates in the olfactory also operates in the visual and auditory systems, these investigators suggested that what was observed in the olfactory system is potentially a mechanism that is shared across many modalities.

The action of the amygdala as a regulator and the function of amygdalar nuclei may explain why deep pressure, such as hugging, may give a person pleasure and light touch may be terrifying to others.

The effect of damage to the amygdala in monkey

The function of the amygdala has been studied using many different methods. One of the earliest was the study of the effect of ablation of the amygdala in animals. The best known of such studies is that reported by Klüver-Bucy [58]. These investigators showed dramatic changes in the behavior of a rhesus monkey after the ablation of the amygdala. An often-cited syndrome is known as the Klüver-Bucy syndrome. This syndrome describes the effect of ablation of the amygdala. The description consists mainly of a lack of social interaction, a lack of fear, and a lowered level of aggression.

However, some of these changes in behavior were most likely caused by the researchers ablating extensive parts of the temporal lobe when they attempted to ablate the amygdala [92, 93]. More recent studies in which the amygdala was ablated bilaterally showed a less pronounced effect on social behavior, but the ablation also made the animals less aggressive or tamer.

Studies by Adolphs and other co-workers [5] have shown that bilateral amygdala damage in humans compromises the recognition of fear in facial expressions while leaving the recognition of facial identity intact [94]. In another study, Adolphs et al. stated "We show here that while bilateral damage to the amygdala impairs recognition of fearful facial expressions, and impairs their recall in imagery, it leaves relatively unaffected verbal knowledge related to fear" [75, 94]. This means that the amygdala may be mainly associated with visual expression such as pictures, images, and more, and that the amygdala is especially suited for processing visual stimuli that communicate significant emotional information such as in facial expressions.

Others studies done by Adolphs and co-workers, that were done in people with damage to the amygdala, "suggest that bilateral, but not unilateral, damage to the amygdala impairs judgments of the intensity of expression of fear, and of expressions normally judged to be very similar to fear, such as surprise" [75].

Furthermore, these investigators stated, "The data showed that bilateral, but not unilateral, damage to the amygdala impairs the processing of some facial expressions of emotion. The recognition of fear appears to be most severely impaired". These investigators summarized their findings as [94]: "Here with the help of one such rare patient, we report findings that suggest the human amygdala is indispensable to:

1. Recognize fear in facial expressions,
2. Recognize multiple emotions in single facial expression; but
3. It is not required for recognizing personal identity from faces. These results suggest that damage restricted to the amygdala causes particular recognition impairments, and thus constrains the notion that the amygdala is involved in emotion.

Effects of surgical removal of the amygdala in a rhesus monkey

The study, published 1937 by Klüver and Bucy, provided some indications about the many functions that are controlled by the amygdala [58]. The study has two important aspects; it was done on primate animals, in contrast to most other studies of the function of the amygdala, and the study was not restricted solely to the role of the amygdala in emotions. Recently, the value of the study has become further evident when the results have been further examined because it has inspired investigators to do more studies in humans.

This old study of the effect of the bilateral removal of the amygdalae (and some other structures of the temporal lobes) by Klüver and Bucy in 1937 [58] showed that the animals, specifically monkeys, exhibited several changes in their behavior such as:

1. "Psychic blindness," an inability to recognize "the emotional importance of events."
2. Showed a lack of fear for items that would ordinarily frighten monkeys
3. Displayed an appetite for improper foods such as rocks or live rats
4. Sought sexual intercourse with unusual partners, including members of other species
5. Became extremely interested in exploring items in their environment
6. Became placid when approached

Studies of the behavioral role of the amygdala in human

The idea that the amygdala is associated explicitly with fear was confirmed in a study of a person with a sporadic disorder, Urbach-Wiethe disease. This disorder is associated with bilateral calcification of the anterior medial temporal lobe and extensively damage the amygdala on both sides. The person with Urbach-Wiethe disease was not able to draw facial expressions of fear, but she could correctly draw facial expressions of happiness, sadness, surprise, anger, and disgust. This deficit in imagining fear was assumed to be the result of a loss of function in the amygdala on both sides [75]. The results of these studies also indicate that the amygdala is closely associated with the visual system.

The first published identification of a human patient with Klüver-Bucy symptoms was made in 1955 by Terzian and Dalle [95]. The signs and symptoms from people in whom the amygdala has been resected or damaged are all similar to those described by Klüver and Bucy [58, 93] in a monkey that had its bilateral amygdala surgically ablated. However, the monkey had some additional symptoms and signs.

Effect of damage to the amygdala in human

Effect of surgical removal of the amygdala nuclei in humans: The Klüver–Bucy syndrome. Dr. Hayman and co-workers in 1998 published the reported the signs of damage to the amygdala in a patient with isolated symmetric damage to the amygdalae and their cortical connections to be:

1. "Psychic blindness."
2. Hyper-sexuality often directed indiscriminately.
3. Altered emotional behavior, particularly placidity.
4. Hyperorality and the ingestion of inappropriate objects (pica).
5. A tendency to react to every visual stimulus.
6. Memory deficits.

Another study in a right-handed patient after left anterior temporal lobectomy for an anaplastic oligodendroglioma had a complete set of Klüver-Bucy symptoms, including psychic blindness, aberrant sexual behavior, hypermetamorphosis, aphasia, and visual agnosia [96].

Lilly et al. (1983) [97] reported on 12 patients with Klüver-Bucy syndrome that occurred after head trauma, during Alzheimer's disease and Pick's disease, and following herpes encephalitis. The symptoms of ablation or damage to the amygdala reported varied in different patients, but the symptoms are different and more complex humans than nonhuman primates. Lilly et al. (1983) [97] have attributed these more complex behavioral syndromes, seen in humans, to the evolutionary advances of the human brain. For example, psychic blindness becomes prosopagnosia or other forms of visual agnosia, hyperorality becomes hyperphagia and bulimia, and hypermetamorphosis becomes distractibility.

Aphasia is a common symptom that has been ascribed to lack of or dysfunction of the amygdala, but in these patients, who had damage to their amygdalae, some of that dysfunction may have been caused by damage to the temporal lobe in general.

The alterations in the emotional behavior in humans include apathy, lethargy, and emotional unresponsiveness. Hypersexuality is rare and copulation, masturbation, and other self-stimulation are uncommon as well, but inappropriate sexual remarks and gestures are frequent. Changes in sexual preference have also been reported. The most commonly reported symptoms are hyperorality and placidity; hypersexuality and sensory agnosia is the least frequently reported [98].

Studies of the effect of damage to the amygdala monkeys and humans

Interpretation of the results of studies of monkeys and humans has implicated the amygdala in the cause of innate and learned fear. It was concluded that the emotional disturbances that were seen in the Klüver–Bucy syndrome, such as the apparent loss of fear, was caused by the lesions of the amygdala and ignoring the damage to neighboring structures.

A comprehensive review of the amygdala was published in the Handbook of Physiology in 1960 by Pierre Gloor [99]. Gloor discussed changes in emotional behaviors after amygdala lesions. He pointed out that most of the aspects of the Klüver–Bucy syndrome, except the visual agnosia, were present after lesions to the amygdala. These aspects are uncritical desires to attend and respond to all environmental stimuli, to ingest non-food objects, or to seek sexual gratification indiscriminately. Gloor wrote: "Obviously, all these behavioral mechanisms are intimately interlocked, and the full syndrome is more than the sum of individual deficits attributable to individual anatomical structures" (see [99] ref. 2, p. 1411).

A series of later studies have found the amygdala to be involved in the origin of innate and learned fear [100, 101]. The investigators of these studies found evidence that the emotional disturbances that were seen in the Klüver–Bucy syndrome, such as the apparent loss of fear, resulted from the amygdala lesions, while the other anomalies depended on damage to adjacent brain structures. The focus on the symptoms that were observed as originating in the amygdala was mainly directed to learned fear. Thus, Pavlovian fear conditioning emerged as the dominant experimental model of the function of the amygdala to study [49] [102].

The amygdala and fear reactions

The connections between sensory systems and the limbic system are the basis for evoking emotional responses to sensory stimuli. The lateral nucleus of the amygdala that receives input from all sensory systems except olfaction, such as the thalamic sensory nuclei and the sensory cortices, may be regarded as the gateway between the dorsal and medial division of the MGB (dMGB and mMGB), the suprageniculate nucleus *(SG)*, and the posterior intralaminar nucleus (PIN) [103].

These polysensory pathways seem to be more important for learned fear than the modality-specific projections to the amygdala from the neocortex because ablation of cortical sensory areas does not prevent conditioning of fear responses to auditory stimuli [103, 104]. Instead, acoustically mediated conditioned fear responses are controlled by the input to the lateral nucleus of the amygdala from the dorso-medial thalamus.

Other functions of the amygdala

In addition to learned fear reactions, the amygdala is also involved in many other behaviorally related functions. For example, the amygdala in human has been shown to be involved in memory [105, 106 1834], dreams, fear, sexuality, and anxiety. It also may be also involved in the ability to make social contacts. The amygdala has also been associated with disorders such as tinnitus [107], pain [108], and developmental disorders such as autism [109] [110] [58].

Structures that connect with the nuclei of the amygdala

The role of the olfactory system

Odors activating the olfactory system can elicit emotions, including fear reactions. The olfactory system has two main parts, the conscious olfactory system and the vomeronasal system.

The conscious part of olfaction

Many different odors can activate the conscious part of the olfactory system, which is responsible for odors eliciting awareness and making it possible to discriminate between many different odors. It is an essential part of what creates the perception of food flavor, which is a combination of olfaction and gustation.

The ascending olfactory pathway reaches parts of the CNS that are different from those of the four other senses. The olfactory pathways mainly project to the allocortex and have abundant connections to the structures of the limbic system, mainly the medial nucleus and the central nucleus of the amygdala (in the monkey [55] [81]). The olfactory pathways project to the uncus and, in humans, probably also the entorhinal cortex, which belongs to the allocortex and is located deep in the temporal lobe. There are extensive connections between the two sides of the olfactory bulbs through the anterior commissure. The anterior commissure is often regarded as a part of the corpus callosum.

It has been questioned whether the olfactory pathways are interrupted by synaptic transmission in the thalamus, as the other sensory systems are, but some investigations have shown that the olfactory fibers indeed make connections with cells in the thalamus. The connections from the olfactory bulb reach many parts of the CNS, but it has no direct projections to the neocortex.

The cells in the olfactory bulb mainly project to the central nucleus of the amygdala (see [1]). This means that the olfactory system has a direct connection to a part of the emotional brain. This explains the role of olfaction in fear and other emotions.

The vomeronasal system

The pathways of the vomeronasal system are different from those of the other or conscious olfactory pathways. The vomeronasal organ sends its signals and information solely to the amygdala, where the axons from the olfactory bulb terminate in the central nucleus of the amygdala.

The vomeronasal system does not have any known cortical projections, and the activation of the system does not produce awareness, but it may affect behavior such as sexual attraction even in humans. The vomeronasal pathways do, however, project to the cortico-medial amygdaloid nucleus of the amygdala [111]. This means that the vomeronasal pathway may be similar to the non-classical pathways of other sensory systems and representing the equivalent of the "low route" to the amygdala that has been described for other sensory systems [67]. The corticomedial nuclei of the amygdala are believed to relay pheromone input from the accessory olfactory bulb to the medial forebrain and the hypothalamus.

The basolateral nucleus also receives input from all other sensory systems via the lateral nucleus of the amygdala. Also, it connects to the central nucleus of the amygdala. The auditory, somatosensory, and visual systems provide input to the basolateral nuclei and the central nuclei through the lateral nucleus of the amygdala. These connects are made through mainly two routes; a direct subcortical short route from the dorsal and medial thalamus or the "low route" [67] [112] and an indirect cortical long route or the "high route" via primary sensory cortices, secondary cortices, and association cortices [55] [112, 113].

The classical sensory system only supply information to the amygdala through the (cortical) "high route". The cortical route, or the "high route", is much longer and carries highly processed information that is subject to interactions from intrinsic brain activity as well as all modalities of sensory information. The "low route" to the amygdala carries "raw" information that is only slightly processed. (To review: the low route has also been labeled "quick and dirty," and the high route has been labeled "slow, but accurate" [114].)

Taste and visceral information can reach the central nucleus of the amygdala through the vagus nerve and, its target nucleus, the nucleus of the solitary tract (NST). Through the parabrachial nucleus, axons from cells in the NST can reach cells in the amygdala. This pathway may make it possible to integrate the olfactory experience with taste and visceral activity, while modulating visceral functions by emotional stimuli.

The abilities of the various connections to activate their targets can, however, not be known from anatomical data. Therefore, the anatomical information alone does not indicate if the connections are functional and how strong the connections are. This requires knowledge about the functional connections. Studies of functional connections in the central nervous system, or "connectivity", is a relatively new subspecialty of neuroscience. It is known from connectivity studies in humans that the strength of functional connections in the brain can vary from time to time and that the functions of connections can be affected by many factors [82, 115].

The cerebral cortex

The role of sensory cortices

If the connections to the auditory cortex pathway are severed basic fear conditioning remains unaltered, but discrimination is altered. This means that the cerebral cortex is not needed for simple fear conditioning. Instead, it allows a person to recognize an object by sight or sound and interpret the environment. It also means that fear conditioning is based on information from the dorsal-medial thalamus, thus the non-classical sensory pathways.

The non-classical pathways from the dorsal-medial sensory thalamus to the amygdala provide only crude information about the outside world, but because they involve only one neural link, they are fast pathways. A fast response may be beneficial in responding to potential danger. A recent study [116] found that neurons in the macaque pulvinar thalamus can respond selectively to the view of snakes in 55 ms, which is likely too short for a cortical route indicating that it must use the low route. Other studies have shown that the amygdala can be activated with latencies from a fear-relevant stimulus as short as 40–120 ms.

The conclusion is that the cerebral cortex is not needed for simple fear conditioning. Therefore, pathways from the sensory thalamus provide only a crude perception of the world. The thalamus—amygdala direct pathway, via the non-classical pathway, also prepares the amygdala to receive more highly processed information from the cortex.

Prefrontal cortex

The prefrontal cortex (PFC) and the hippocampus both have extensive connections to the amygdala, which is essential for conditioned fear and associative emotional learning. The PFC is thought to be responsible for reactivating past emotional associations and is decreased in both responsiveness and density.

Also, there are many connections between nerve cells in the central nucleus of the amygdala and nerve cells in the prefrontal cortex where decision-making is mainly located. These connections are "reciprocal", as are most connections in the brain, meaning that the amygdala can influence the function of the prefrontal cortex and the activity in the prefrontal cortex can influence the amygdala. Connections to the amygdala make it possible to modulate already learned behaviors, and behavior can then modulate the activity in the central nucleus of the amygdala.

Orbitofrontal cortex

The orbitofrontal cortex is another part of the frontal lobe that receives input from the amygdala. The cells in the orbitofrontal cortex send connections back to the amygdala, and its cells have direct and indirect connections to the hypothalamus.

The orbitofrontal cortex is also known as the ventromedial prefrontal cortex. The orbitofrontal cortex receives input from the mediodorsal thalamus, and it is assumed to be involved in emotions and reward in decision making. Most of the present knowledge about this part of the brain comes from animal studies, but little is known about the connections to other parts of the brain. Ultimately, the function of the human orbitofrontal cortex is poorly understood.

The anterior cingulate cortex and the basal ganglia

The anterior cingulate cortex (ACC) has a central role in processing emotional experiences at the conscious level. The anterior cingulate cortex is coupled with limbic structures, like the amygdala, and the basal ganglia structures such as the putamen, pallidum, and caudate nucleus. Emotionally related learning is mediated through the interactions of the basolateral amygdala and hippocampal formation. Also, motivational responses are processed through the dorsolateral prefrontal cortex. The anterior cingulate cortex (ACC) is involved in the down-regulation of (reducing) emotional distress that can be achieved through various cognitive control strategies. This occurs through the inhibitory effect that ACC has on the amygdala, which is also the basis for emotional reactivity. In people with depression or risk for depression, this inhibitory effect of the ACC on amygdala activation is reduced.

Sensory pathways to visceral structures

Figure 3.19 shows a contemporary description of the pathways that emotional signals from the environment can take to reach visceral structures of the brain.

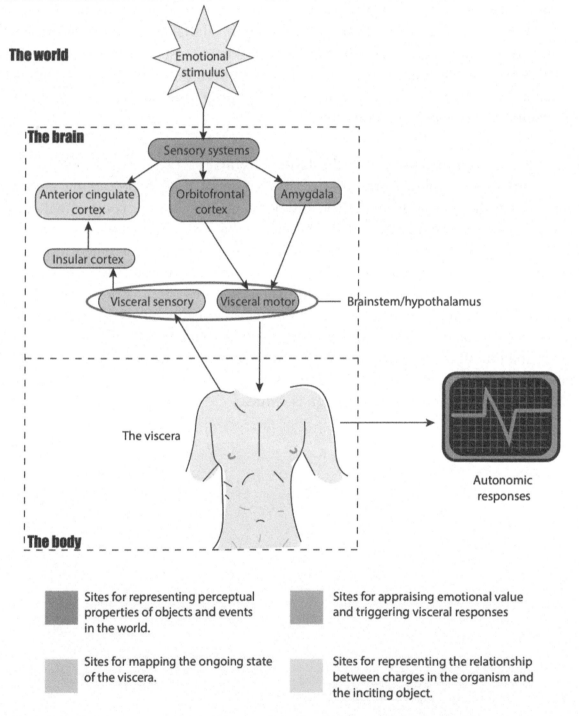

Figure 3.19 A current description of how emotional, sensory signals can reach visceral parts of the brain. Only the visceral response is represented, although emotion includes endocrine and somatomotor responses as well. (Based on: Bechara and Naqvi, 2004.) [117] (Artwork by Liliana Cabrera.)

This figure shows that information, which can be derived from the environment or recalled from memory, is made available to the amygdala and orbitofrontal cortex. The sites of emotion execution include the hypothalamus, the basal forebrain, and nuclei in the brainstem tegmentum. Visceral sensations reach the anterior insular cortex by passing through the brainstem. Feelings result from the re-representation of changes in the viscera about the object or event that initiated them. Additionally, the anterior cingulate cortex is a site of this second-order mapping.

The hypothalamus

The hypothalamus is activated by the amygdala, and it acts as the effector organ of the amygdala. The hypothalamus controls many kinds of actions such as the secretion of chemical factors from the pituitary gland. Those chemical factors, in turn, control the liberation of several different molecules from the adrenals, such as the stress hormones, cortisol, and adrenergic hormones such as epinephrine.

Figure 3.20 and 3.21 show the main afferent and efferent connections to and from the hypothalamus.

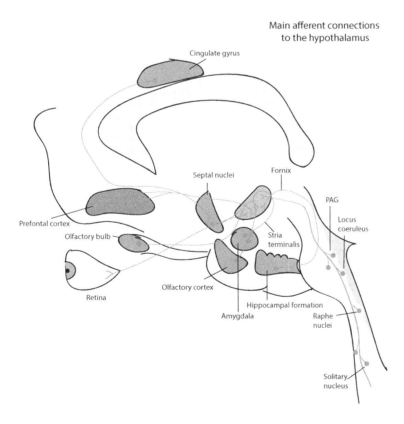

Figure 3.20 The main afferent connections to and from the hypothalamus. (Artwork by Liliana Cabrera.)

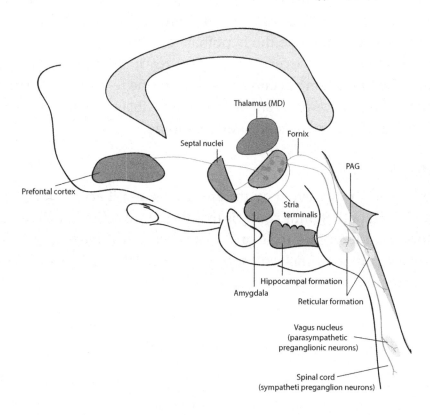

Figure 3.21 Main connections to (A) and from (B) the hypothalamus. (Artwork by Liliana Cabrera.)

Activation of the hypothalamus-pituitary-adrenal (HPA) axis from the amygdala causes many of the signs and symptoms of emotions. The hypothalamus causes the "fight or flight" reaction. Emotions, including fear, can activate the hypothalamus. It is also activated through other routes such as those of the autonomic nervous system and stressful situations.

The hippocampus

The hippocampus is another structure that is very much involved in the expression of emotions. The hippocampus is a complex structure that is mainly associated with memory, but it also has other functions, and it connects and interacts with many other systems of the brain. The hippocampus plays an essential role in memory. The amygdala and the hippocampus belong to the old part of the brain that can be found in many kinds of mammals, including rats and mice and even in reptiles.

The dorsal sensory stream reaches the parahippocampal cortex and the ventral stream converge with the output of the parahippocampus into the entorhinal cortex, which is then the input and output pathway of the hippocampus proper, consisting of the dentate gyrus (DG) and areas of "Ammon's horn" (cornu ammonis, CA) -- CA3 and CA1. CA3 represents the primary "engram" for the episodic memory, while CA1 is an invertible encoding of EC, such that subsequent recall of the CA3 engram can activate CA1 and then EC, to reactivate the full episodic memory out into the cortex. All connections are bidirectional.

The hippocampus sits on "top" of the cortical hierarchy and can encode information from all over the brain, binding it together into episodic memory (Figure 3.22).

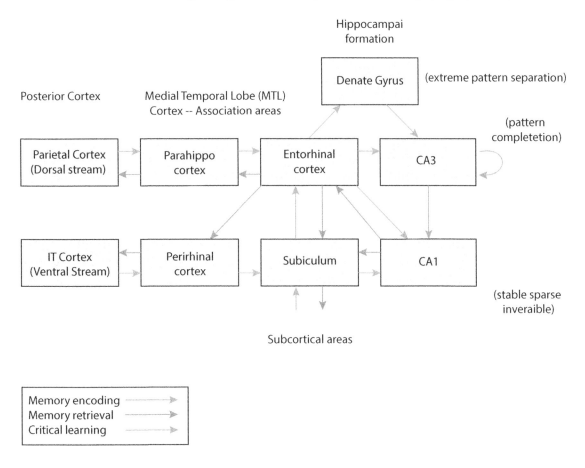

Figure 3.22 The functional aspects of the hippocampus and its connections with other structures of the brain regarding memory and learning. (Based on CCN Book/Brain Areas.) (Artwork by Liliana Cabrera.)

Learning usually requires repeated exposure, but the effective learning that occurs in connections with fear often only depends on a single event. This form of learning seems to depend on the areas of the medial temporal lobe (MTL), including the perirhinal cortex. Another form of one trial behavioral learning involves mechanisms that support active maintenance of memories in an attractor state or a state of working memory in the prefrontal cortex. This form of memory does not require a weight change at all but can nevertheless rapidly influence behavioral performance from one instance to the next.

Connections of vagal afferents

Activity in the ascending part of the vagus nerve can cause cortical arousal and the facilitation of plastic changes because these axons terminate on cells in the nucleus of the solitary tract (NST). The NST then projects to many areas of the CNS, which provides arousal and the facilitation of neural plasticity particularly. Also, the vagus nerve innervates receptors in the viscera, which thereby can influence the excitability of cortical cells and promote plastic changes. This is similar to what the activity in the nucleus of Meynert can do. The amygdala modulates many functions controlling arousal through the vagus nerve and the nucleus of Meynert.

Figure 3.23 shows some of the structures that can be reached by signals in the ascending part of the vagus nerve. The terminals of the vagus nerve synapse directly within the NST, which conveys information from organs in the abdomen, the heart, and lungs to structures that are involved in memory functions such as the amygdala, hippocampus, and frontal cortex via a polysynaptic pathway to the locus coeruleus (LC). Norepinephrine (NE) is one of the primary transmitters to mediate synaptic communication between these structures.

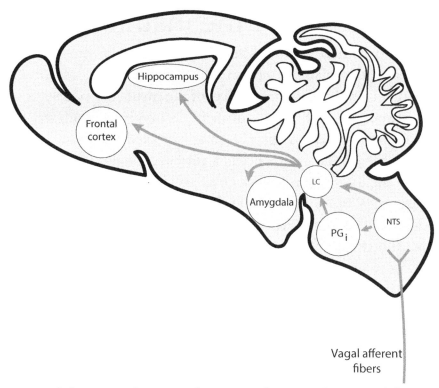

Figure 3.23 Schematic diagram depicting the contribution of the nucleus of the solitary tract (NST) as both a recipient of peripheral inputs from the vagus nerve and a transmitter of these visceral signals to limbic structures that process memory after emotionally arousing events. (Based on: McIntyre, McGaugh, and Williams 2011.) [105] (Artwork by Liliana Cabrera.)

The motor portion of the vagus nerve provides the autonomic or parasympathetic control of organs in the abdomen, the heart, and the lungs.

The PAG and the freezing reaction

Studies in rodents have shown that freezing depends on amygdala projections to the periaqueductal gray (PAG). In humans, freezing-like behaviors are implicated in the development and maintenance of psychopathology, but neural mechanisms underlying freezing or its distinctive autonomic response profile have not been identified [118].

The role of the insular lobe

The insular lobe is a little-known structure that may be the site of unexplained symptoms. The insula lobe, also known as the fifth brain lobe, is essential for emotions and fear. Unfortunately, the functional aspects of the insula are poorly known. In 2006, Paulus and Stein proposed that the anterior insula is involved in anxiety by augmented detection of the difference between observed and expected body state [119]. Regarding functional anatomy, the insular cortex holds a primary position in interoception and is thought to mediate the integration and associative learning that underlies higher level interoceptive inference [120].

The insular cortex is often the site of epileptic foci. Only recently has this part of the brain been studied functionally. Some of these studies were byproducts of the diagnostic workup that was developed for patients with severe epilepsy [121]. Also, localizing epileptic foci is critical to the surgical treatment of severe epilepsy cases that do not respond to other forms of treatment. In such patients, new methods for accurate localization of the anatomical area where the epileptic seizures started can be done by electrical stimulation and recordings using electrode arrays that are temporarily implanted in the insula. These diagnostic methods made it possible to study the neural activity in different regions of the insula, and the perception of stimulation of different regions of the insula could be studied in awake cooperating patients in connection with the workup of such patients [121]. These opportunities have provided a wealth of information about some of the functions of the insula [122].

The results of these studies indicate that the insular lobe, which is located under the temporal and parietal lobes, is involved in sensory functions related to taste and sensations from the digestive organs. A new understanding of the functions of the insula comes from the use of a diagnostic method in neurosurgery. That method has made it possible to study the function of the insular lobe in humans. From such studies, much has been learned about the normal function of the insular lobe [121, 122].

It is now believed that the insular cortex has a primary role in interoception, and the insular cortex is thought to mediate the integration and associative learning that underlies higher level interoceptive inference [120]. The prefrontal cortex (e.g., anterior cingulate and ventromedial prefrontal region) integrates the ensuing representations as part of hierarchical inference which underlies emotional awareness, where the brain works as a prediction machine [123].

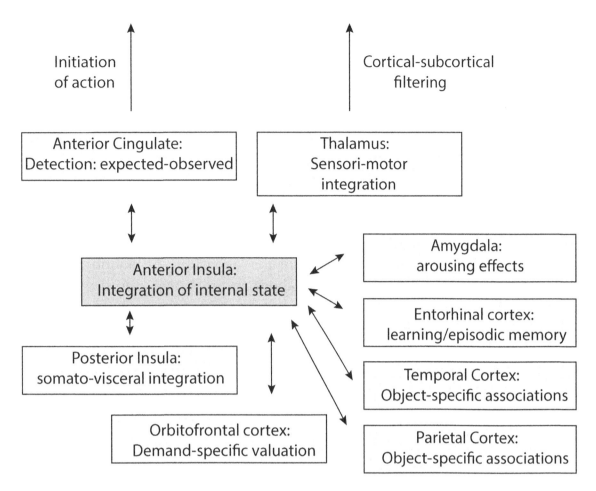

Figure 3.24 Schematic diagram depicting the contribution of the nucleus of the solitary tract (NST) as both a recipient of peripheral inputs from the vagus nerve and a transmitter of these visceral signals to limbic structures that process memory after emotionally arousing events. (Based on: McIntyre, McGaugh, and Williams 2011.) [105] (Artwork by Liliana Cabrera.)

This figure shows the extensive connections to and from the insular lobe. The connections shown are the anatomical connections. Additionally, much less is known about the functional connections and the physiological implications of these connections.

The cerebellum

In 2015, Strata showed that the cerebellum, with its connections to the autonomic nervous system, plays a vital role in the expression of emotions such as fear [124]. Studies in rodents have shown that reversible inactivation of the cerebellar vermis during the consolidation or the re-consolidation period hampers the retention of the fear memory trace. It was shown that there is a long-term potentiation of the excitatory synapses between the parallel fibers and the Purkinje cells. There is also long-term potentiation of the feed-forward inhibition mediated by molecular layer interneurons [124, 125].

In humans, the cerebellar hemispheres are also involved at a higher emotional level. Imaging studies in humans show the cerebellum is activated during mental recall of personal, emotional episodes and learning of a conditioned or unconditioned association involving emotions.

The cerebellar vermis participates in fear learning and memory mechanisms related to the expression of autonomic and motor responses of emotions. The importance of these findings is evident when considering the cerebellar malfunctioning in psychiatric diseases, such as autism and schizophrenia, which are characterized behaviorally by impairments of the processing of emotion-related information [124].

Cognitive symptoms, emotional changes, and autonomic changes have also been reported as consequences of cerebellar lesions [126, 127]. These changes and symptoms support the hypothesis of the involvement of the cerebellum in emotions. There is substantial evidence for cerebellar abnormalities in emotional disorders, including schizophrenia and depression [126].

The role of the interpositus nucleus in autonomic and emotional functions has achieved a broad consensus among scientists [128]. The interpositus nucleus is located deep in the cerebellum. This nucleus receives its afferent supply from the anterior lobe of the cerebellum and sends output via the superior cerebellar peduncle to the red nucleus.

Emotions and memory

Memory is essential to numerous aspects of fear and other emotions. Memory plays a vital role in fear conditioning, which is essential for survival because it makes it possible to store information about events and objects that represent severe risks to an animal.

Memory is also important for humans for remembering what events or objects to actually be afraid of. Also, the memories of fearful events are the basis for mood disorders such as PTSD [129]. This section describes some basic properties of memory and how they relate to fear, anxiety, and some specific mood disorders.

The basolateral and central nuclei of the amygdala are involved, in a complicated way, in memory consolidation through interactions between many different systems [105]. McGaugh and his colleagues have shown that learning emotional events depends on adrenergic substances. They also have shown that learning is impaired after administering β-adrenergic blocker such as propranolol [106]. Stimulation that reaches the basolateral nucleus of the amygdala, either from the thalamus (the subcortical or low route) or the cerebral cortex (the cortical or high route), can change synaptic efficacy as an expression of "learning". Such "learning" may also be regarded as neural plasticity. Neural plasticity can "switch on" neural circuits that are not normally activated.

Adrenergic substances may affect memory consolidation

Adrenergic substances are essential for memory consolidation. In different studies, recall was shown where a beta-adrenergic blocker, propranolol, was used to reduce the effect of the administration of an adrenergic substance [130].

The level of adrenergic substances in the blood increases in situations of trauma or other exciting events. That increase in adrenergic substance levels affects many systems in the brain and body. Drugs, such as propranolol, reduce the effect of adrenergic substances, like catecholamines such as adrenaline and noradrenaline, because they block beta-adrenergic receptors. The effect of propranolol, one of the first beta-adrenergic blockers created, on fear is related partly to reducing memory consolidation [130], including the memory of fear matters [131].

Retrieval of learned information or "remembering" is facilitated by adrenergic substances. The relationship between effectiveness in the retrieval of memory and the concentration of the adrenergic substance is an inverted U-shaped curve. This curve indicates that the effectiveness of retrieval increases with increased adrenergic concentration (increased stress) to reach the maximal effect. After achieving the maximal effect, increases in the adrenergic concentration impair retrieval.

Yerkes and Dodson first documented the inverted U-effect of stress on performance in 1908 [132]. Also, they showed the inverted U-function between arousal and performance. While adrenergic substances enhance memory consolidation [105, 106], it is remarkable that the same dose of glucocorticoids, another stress hormone that enhances consolidation of memory, can impair retrieval of memory [133].

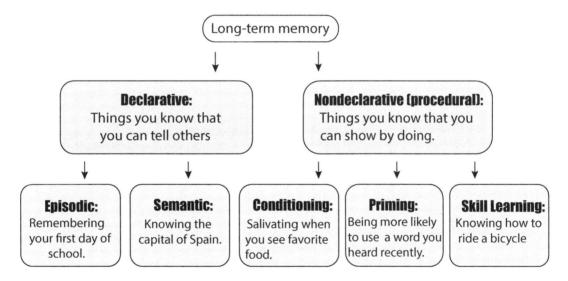

Figure 3.25 Types of long-term memory. (Adapted from Biological Psychology 6e) (Artwork by Liliana Cabrera).

Neural Circuitry of PTSD

Many neural circuits and structures can modulate fear, including the amygdala, PFC, and hippocampal regions. The hippocampus is thought to play a role in explicit memories of traumatic events and in mediating learned responses to contextual cues. In post-traumatic stress disorder (PTSD), the hippocampus decreases in volume and the responsiveness to traumatic stimuli is increased [129]. The connections from the amygdala to the hippocampus emphasize the importance of memory in emotions including fear. Sensory stimuli can also reach the hippocampus, which then affects the consolidation of memories related to emotional experiences.

Synaptic plasticity is the underlying basis of learning and memory, and maladaptive neuroplasticity plays a role in many diseases [134], including PTSD [84, 129]. Behaviorally, people with PTSD have shown an increased sensitization to stress, the overgeneralization of fear associations, and a failure to extinguish learned fear (Figure 3.26). Persons who demonstrate resilience to PTSD and who recover from traumatic experiences, can discriminate between fearful and non-fearful stimuli, as well as display the normal extinction of fear memories (Figure 3.26).

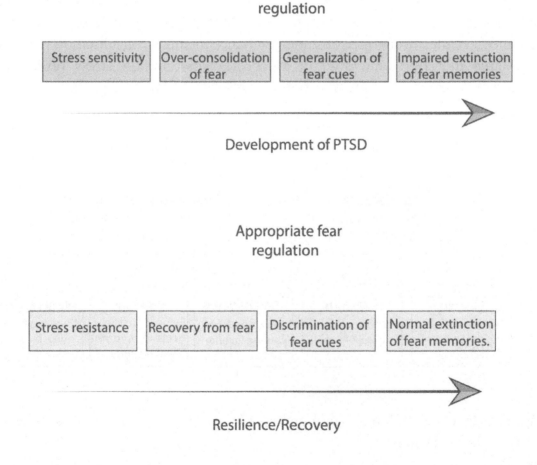

Figure 3.26 Fear regulation. (Based on Mahan and Ressler, 2012, [135].) (Artwork by Liliana Cabrera).

Emotional memory is a particular form of memory that is the raw recollection of information that is related to the happening of something significant. Emotional memories are stored in the amygdala thus below the level of consciousness. Emotional memories are retrieved independently and perceived differently.

The amygdala can modulate memory

Several studies have demonstrated that emotions enhance memory encoding and facilitate later recall. This effect has been ascribed to the increase in the secretion of an adrenergic substance that occurs in situations such as trauma [130].

The amygdala also triggers the release of stress hormones by way of the hypothalamic–pituitary–adrenal (HPA) axis [15], which feeds back onto memory consolidation and storage sites. Also, the HPA axis feeds back to the amygdala itself to enhance memory over longer time intervals [136]. Recent studies have revealed that the amygdala plays a vital role in the modulation and activation of memory functions. Additionally, the amygdala can enhance memory consolidation by facilitating neural plasticity and information storage processes in parts of the brain that receive signals from the amygdala [137]. Recent in studies in animals, as well as studies in humans, have identified the most important of these brain regions and neurochemical processes involved using neuropsychological and neuroimaging techniques.

Potential mechanisms by which the amygdala mediates its influence on memory

Emotional learning takes place intrinsically in the amygdala. Direct and indirect neural projections from the amygdala target several memory systems in the brain, including those that serve working memory, declarative memory, and various non-declarative forms of memory. For example, procedural learning, priming, and reflexive conditioning are some of the forms. Consciousness is necessary for experiencing fear and other emotions. Consciousness is complicated, but one form of consciousness is related to what makes up the working memory [138].

The many different structures that receive input from the amygdala makes it possible for emotional stimuli to achieve many different reactions. Some of these functions are illustrated in Figure 3.27, where solid arrows indicate direct connections and dashed arrows indicate indirect connections.

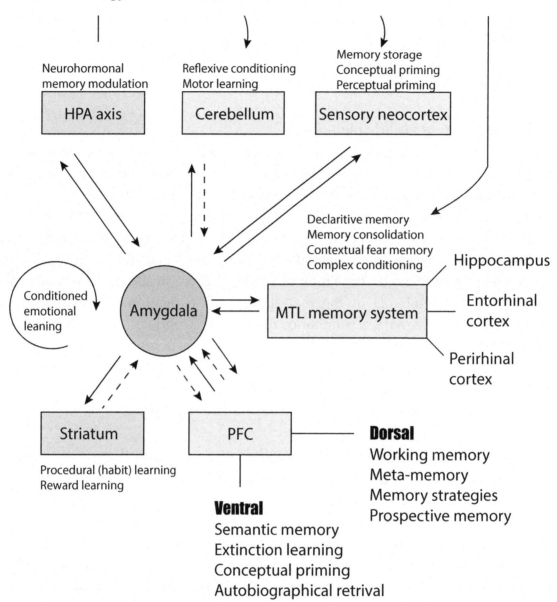

Figure 3.27 Potential mechanisms by which the amygdala mediates an influence of the emotional arousal on memory. Solid arrows indicate direct connections; dashed arrows indicate indirect connections. Blue labels indicate connections with subcortical structures. MTL, medial temporal lobe; PFC, the prefrontal cortex. (Based on: LaBar and R. Cabeza, 2006.) [136]. (Artwork by Liliana Cabrera.)

Complex conditioning refers to various higher-order conditioning procedures that are hippocampal-dependent, including trace conditioning and conditional discrimination learning.

To review, the amygdala also triggers the release of stress hormones by way of the hypothalamic–pituitary–adrenal (HPA) axis, which feeds back onto memory consolidation and storage sites, as well as the amygdala itself to enhance memory over longer time intervals (Figure 3.28).

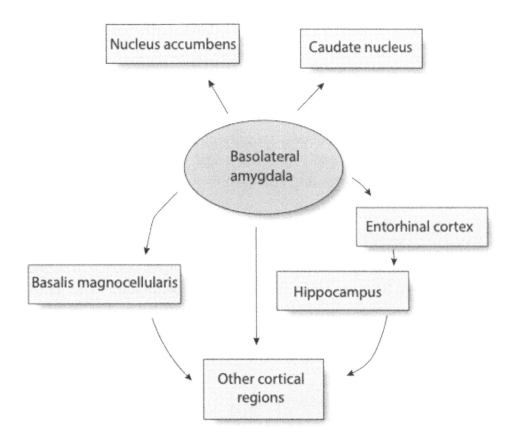

Figure 3.28 Anatomical descriptions of the connections from the basolateral nucleus of the amygdala emphasizing the participation of amygdala nuclei in many of the networks that are involved in memory consolidation. (Based on McGaugh, 2002, [139].) (Artwork by Liliana Cabrera.)

Solid arrows indicate direct connections; dashed arrows indicate indirect connections. Blue labels indicate connections with subcortical structures. MTL, medial temporal lobe; PFC, prefrontal cortex [136].

Feedback loops between pain, emotions, and cognition
There are many examples of functional connections that form loops. Such connections have significant influences on emotions, including fear. The functional connections are subjected to the changes in strength brought about by the activation of neuroplasticity, by substances secreted by structures in the body, and from drugs that are administered for various purposes.

The emotional motor system

The emotional motor system (EMS) can activate a wide range of structures resulting in vigilance, attention, and pathological gastrointestinal symptoms. The EMS can cause the activation of the neuroendocrine system along with autonomic and sensory modulation. Also, the EMS can be activated by psychosocial or exteroceptive and interoceptive stressors, autonomic, pain modulatory, and neuroendocrine responses.

The organism's response to stress is generated by a network comprised of integrative brain structures, such as the subregions of the hypothalamus (paraventricular nucleus, PVN), the amygdala, and the periaqueductal grey particularly. These structures receive input from visceral afferents, somatic afferents, and cortical structures. Cortical inputs include the medial prefrontal cortex and subregions of the anterior cingulate and insular cortices.

The primary descending outputs of the EMS to the periphery are autonomic, pain modulatory, and neuroendocrine responses. Ascending outputs of the EMS to the forebrain include attentional and emotional modulation and feedback from the gut to the emotional motor system. This occurs in the form of neuroendocrine (adrenaline, cortisol) as well as visceral afferent mechanisms mediated by the vagus nerve. Critical ascending outputs to the forebrain include attentional and emotional modulation [140][144].

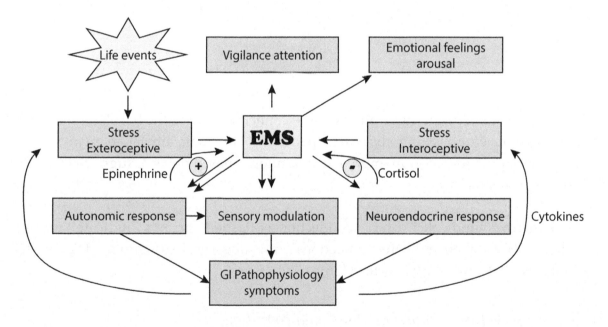

Figure 3.29 Emotional motor system (EMS) pathways. (Based on: Mayer, 2000.) [141] (Artwork by Liliana Cabrera.)

The emotional body language (EBL)

Body language is essential to emotions and, in the visual sense, plays an essential role in communicating emotions.

Three interrelated brain networks involved in emotional body language (EBL)

Three interrelated brain networks are involved in the emotional body language (EBL). The components of the EBL are the reflex-like EBL involving the superior colliculus, pulvinar thalamus, striatum, and the amygdala which cause body awareness through the activation of the insula, somatosensory cortex, anterior cingulate cortex, and the ventromedial prefrontal cortex. The EBL consists of a reflex-like EBL and the visuomotor perception. Emotions in connection with the sense of vision play essential roles in communicating emotions.

Body awareness is produced by the reflex-like EBL and the visuomotor perception of EBL.

Reflex-like emotional body language (EBL) (orange) involves the superior colliculus (SC), pulvinar (Pulv), striatum and amygdala (AMG). Body awareness of EBL (green) involves the insula, somatosensory (SS) cortex, anterior cingulate cortex (ACC) and ventromedial prefrontal cortex (VMPFC). Visuomotor perception of EBL (blue) involves the lateral occipital complex (LOC), superior temporal sulcus (STS), intraparietal sulcus (IPS), fusiform gyrus (FG), amygdala (AMG) and premotor cortex (PM).

Visual information from EBL enters in parallel via a subcortical (red) and a cortical (blue) input system. Feedforward connections from the subcortical to the cortical system and body awareness system are shown in red, reciprocal interactions between the cortical network, and body awareness system is shown in blue [138].

Other structures connecting to the emotional brain

Involvement of the autonomic nervous system

The ample connections from the amygdala to the autonomic nervous system can explain why, for instance, fear causes autonomic reactions such as palpation, sweating, and redirection of blood flow from viscera to muscle. These reactions are also known as the fight or flight reaction. Fear also causes increased muscle tone and may paradoxically cause increased visceral motility. These reactions may be mediated by the connections to the motor nuclei of the vagus nerve. Other strong emotional reactions to sensory input may cause fainting spells because of paradoxical or parasympathetic responses from the cardiovascular system. Seeing blood, particularly one's own blood, is a potent stimulus for such reactions. Medical students' first-time visit to the operating room can cause fainting spells.

The central and medial nuclei of the amygdala are involved in autonomic functions, whereas the basolateral nuclei are involved in conscious functions that are related to the frontal and the temporal lobes [55, 92]. Also, the basolateral nuclei are involved in numerous other functions, such as providing arousal and promoting neural plasticity (through the nucleus of Meynert) [67, 142].

The autonomic nervous system involves both sensory and motor systems. It receives input from the central nervous system through the peripheral nervous system. Additionally, the autonomic nervous system can affect the function of systems of emotions, including fear.

The autonomic nervous system is discussed in more detail under "Autonomic nervous system".

Hierarchical organization of homeostatic reflex systems within the brain-gut axis

Emotional stimuli can activate the visceral system through three different routes, one using the amygdala, another using orbitofrontal cortex, and a third route using the anterior cingulate cortex and insular cortex (Figure 3.30). Activation of the visceral system produces different kinds of autonomic responses.

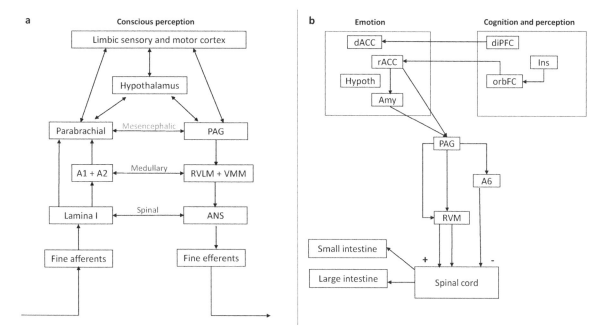

Figure 3.30 Hierarchical organization of homeostatic reflex systems within the brain-gut axis.

(a) Homeostatic afferents that report the physiological condition of the GI tract terminate in lamina I of the dorsal horn (corresponding vagal afferent pathways are not shown). The ascending projections of these neurons provide the basis for autonomic reflex arcs at different levels of the brain-gut axis (enteric nervous system reflexes not shown).

(b) Cortical modulation of homeostatic afferent input to the central nervous system. Prefrontal regions modulate activity in limbic and paralimbic regions, subregions of the anterior cingulate cortex, and hypothalamus, which in turn regulate the activity of descending inhibitory and faciliatory descending pathways through the periaqueductal gray and pontomedullary nuclei. Activity in these corticolimbic pontine networks mediates the effect of cognitions and emotions on the perception of homeostatic feelings, including visceral pain and discomfort.

Abbreviations: ANS, autonomic nervous system; dlPFC, dorsolateral PFC; orbFC, orbitofrontal cortex; PAG, periaqueductal gray; RVM, rostroventral medulla; RVLM, rostroventrolateral medulla; VMM, ventromedial medulla; A6, locus coeruleus; Ins, insula; Hypoth, hypothalamus; Amy, amygdala; ACC, anterior cingulate cortex. (Based on Mayer, and Tillisch, 2011.) [143] (Artwork by Sandra Akioto.)

Limbic, paralimbic, and prefrontal centers provide modulatory influences on the gain of these reflexes. The interoceptive input into these reflexes is generally not consciously perceived except in pathological conditions or in situations that require an action, such as in the presence of pain.

Cortical modulation of homeostatic afferent input to the central nervous system

Prefrontal regions modulate activity in the limbic and paralimbic regions, subregions of the anterior cingulate cortex, and the hypothalamus, which in turn regulate the activity of the descending inhibitory and facilitatory pathways via the periaqueductal gray and pontomedullary nuclei. Activity in these corticolimbic pontine networks mediates the effect of cognition and emotions on the perception of homeostatic feelings, including visceral pain and discomfort [143] (Figure 3.31).

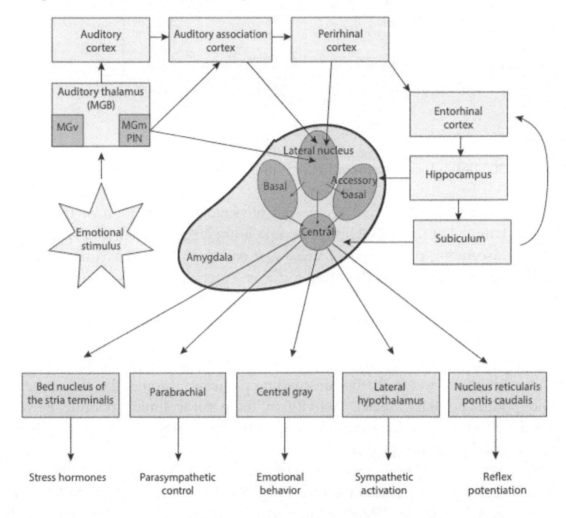

Figure 3.31 Expression of different emotional responses by the amygdala. (Based on Limbic System: Amygdala, Anthony Wright, Ph.D., Department of Neurobiology and Anatomy, The UT Medical School at Houston). (Artwork by Liliana Cabrera.)

It is seen from Figure 3.31 that an emotional stimulus (sound) can reach the lateral nucleus of the amygdala in two routes. The routes are either the short route directly from the dorso-medial thalamus or through a long route from the ventral auditory thalamus. The long route consists of primary, secondary, and association cortices (non-classical and classical routes, respectively).

There is also an even longer route that carries sensory (auditory) information further through the entorhinal cortex and the hippocampus. From the entorhinal cortex, the information travels to unspecific parts of the amygdala. From the hippocampus, the data can reach the amygdala after going to the subiculum. The outputs from the amygdala go from the central nucleus, from where there are direct connections to the bed nucleus of the stria terminalis and from where stress hormones are sent, to other structures. Sympathetic activation from the lateral hypothalamus is another output channel from the central nucleus of the amygdala.

The parabrachial is another output from the central nucleus that projects to the parasympathetic system. Nucleus reticularis pontis caudalis receives its input from the central nucleus and provides reflex potentiation. The periaqueductal gray (PAG) also known as the central gray, controls emotional behavior.

Pathways from the thalamus to the amygdala are particularly crucial in emotional learning. Output pathways from the central nucleus of the amygdala make extensive connections with the brain stem for emotional responses. The central nucleus of the amygdala also makes extensive connections with cortical areas through the nucleus basalis. Cholinergic projections from the nucleus basalis (the nucleus of Meynert) to the cortex are thought to arouse the cortex. The nuclei of the amygdala, sensory cortices, and orbitofrontal cortex interact.

Figure 3.32 shows functional connections between the cells in the nuclei of the amygdala and other brain structures.

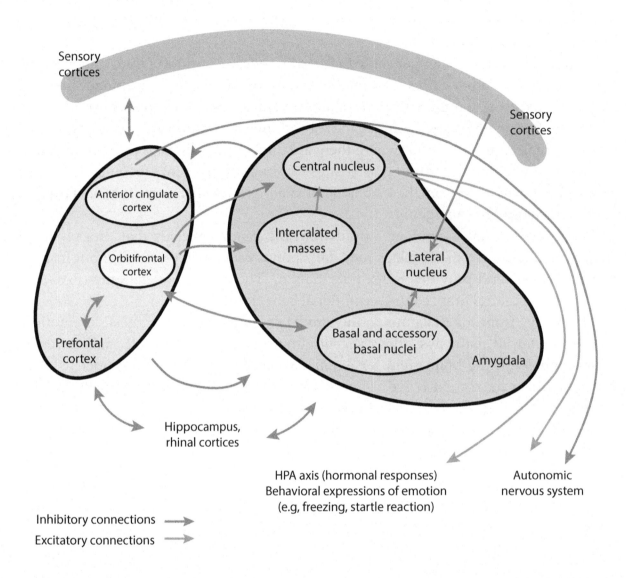

Figure 3.32 Connectivity to and from the macaque amygdala showing the multiple subdivisions and the extensive connections with the frontal lobe. Cells in its multiple subdivisions make extensive connections with cells in many parts of the frontal lobe. Blue arrows: Inhibitory connections, Red arrows (different thickness): Excitatory connections. (Original work from Malisch and Garland, 2004 [144].) (Based on Salzman and Fusi, 2010.) [145] (Artwork by Liliana Cabrera.)

The lateral nucleus of the amygdala receives input from the sensory cortices, and the output from the nucleus is the central nucleus and the autonomic system that is inhibitory. The target of this output is the HPA axis. There is also output from the basal nuclei to the cerebral sensory cortices and the orbitofrontal cortex. The orbitofrontal cortex also connects to the intercalated masses within the amygdala and the central nucleus. Cells in the anterior cingulate cortex make a two-way connection to the sensory cortices.

The prefrontal cortex receives and sends information to the orbitofrontal cortex. Also, the hippocampus and rhinal cortices have two-way communication with the prefrontal cortex and the amygdala. Red arrows (different thickness): Excitatory connections [145].

It is also seen that the hypothalamic pituitary adrenal (HPA) axis is involved in the human stress response. Fear-signaling impulses in cells in the central nucleus of the amygdala activate the sympathetic nervous system and modulate the function of the different components of the HPA axis (Figure 3.33).

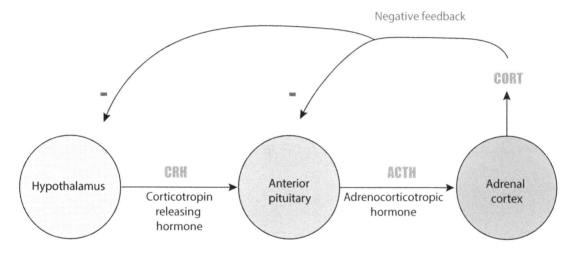

Figure 3.33 The HPA axis. (Artwork by Liliana Cabrera.)

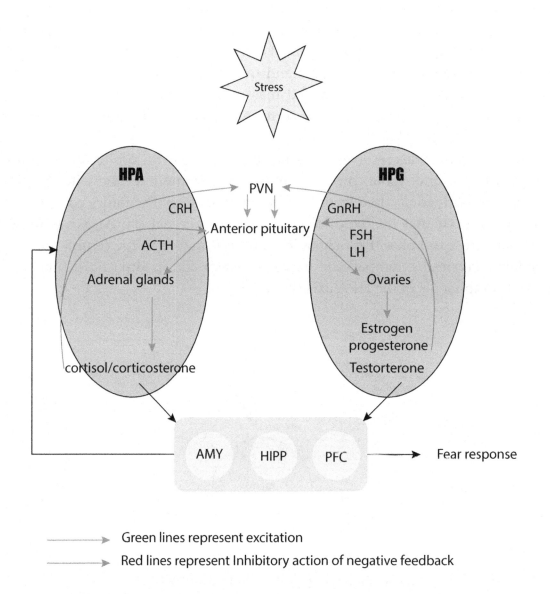

Figure 3.34 The hypothalamic–pituitary–adrenal (HPA) axis [15] vs., the hypothalamic–pituitary–gonadal (HPG) axis. Red lines represent the inhibitory action of negative feedback; green lines represent excitation. (Based on Maeng and Milad, 2015.) [146] (Artwork by Liliana Cabrera.)

The sequence of events causing rage of the brain

Extremely negative experiences can start a series of events in many structures of the brain that begins in the amygdala and cause a rage of the brain. Then, the cardiovascular and the endocrine system activate causing what is known as rage in the brain. Additionally, fear can reduce serotonin levels in the brain and promote aggression.

Rage involves an innate emotional system in the human brain that contributes to the survival of species. In animals, stimulating specific neurocircuitry evokes rage in the brain. Physical brain damage can make these rage neurocircuitry more responsive, and evidence indicates that extremely negative experiences may result in physiological changes in the brain that may predispose an animal, including humans, to bouts of rage.

Applying electrical stimulation to "slightly different brain zones" in laboratory animals evokes three distinct kinds of aggression. 1) predatory aggression, 2) rage-like aggression, and 3) inter-male aggression or dominance aggression. Panksepp (1998) points out, "prolonged social isolation or hunger may increase all forms of aggression, while high brain serotonin activity may reduce them all." [147]

Rage in the brain can be alerted by a stimulus from the environment and then placing the brain in an emotional mode from a thinking mode. An external (sensory) stimulus can initiate a sequence of events in the logic center of the primitive (old) brain that trigger release of hormones that are related to emergencies such as those of an attack on the body. These hormones elicit many changes in the basic functions of the brain especially the cardiovascular system but also the digestive system and the immune systems are affected.

The emotional model is a primitive state where transmitters such as adrenaline and testosterone are in control of many functions [147]. This process is eliciting more hormones by positive feedback. After this series of events the brain is slowly returning to its normal thinking mode and it takes approximately 20 minutes to change from this emotional mode back to a thinking mode.

Sequence of events causing a rage in the brain

1 Brain (amygdala) alerted by external stimuli

6 Take (average) 20 minutes to move from the emotional to thinking brain again.

2 By passes the logic centre and goes straight to the primitive brain

Rage in the brain!

5 As blood pumps, more hormones produced creating more anger/rage

3 Brain triggers adreniline & testosterone hormones "under attack"

4 Blood pumps faster, increased cardiovascular talk louder & quicker!

Figure 3.35 Rage in the brain. (Based on: Kinesiology Source, Bayside Melbourne Victoria, Australia.) (Artwork by Liliana Cabrera.)

Modulation of emotions

Many factors can modulate the neural responses to emotions. One example is the nuclei of the amygdala, which can influence the early neural processing of sensory stimuli as well as the behavioral response to those stimuli [87]. The investigators, in a study of the olfactory system, showed how the nuclei of the amygdala becomes critical for emotional learning, valence coding, and stress. As a result, the amygdala shapes sensory input to the brain through their connections to the locus coeruleus.

Sensory gating is an example of such interaction because inactivating the central, basolateral, and lateral nuclei of the amygdala selectively strengthened olfactory inputs to the brain. This linkage of amygdala and locus coeruleus (LC) output to primary sensory signaling may be involved in creating affective disorders, including sensory dysfunctions like hypervigilance, attentional bias, and impaired sensory gating.

Emotions can affect the immune system

The list of diseases where the immune system plays an important role is increasing. The effect of emotions on the immune system is similar to that from stress, but knowledge about the details on this process is lacking. However, it is known that stress can affect inflammatory processes and cancer. How to reverse the effect on the immune system is now a target of many studies [148] [149].

The role of the immune system

Earlier in this book, it was emphasized that maladaptive neuroplasticity plays an essential role in severe tinnitus, spasticity, and chronic neuropathic pain. Recent research now includes fibromyalgia and several other diseases in the group where the immune system plays an important role. Understanding which factors are implicated in the causes of the symptoms of these diseases has also grown, and BDNF is now included in addition to maladaptive plasticity.

These factors can be affected by relatively simple means which is a favorable scientific development. BDNF can be increased by physical exercise and through the administration of Omega 3. It is to be noted that a low ratio between Omega 6 and Omega 3 is important. The ration can easily be decreased by the intake of Omega 3, which has no known side effects.

Immune receptor cells in the upper part of the small intestine affect the function of many parts of the nervous system [150]. Recent studies have shown some of the critical neural roles of the immune system. The immune system affects functions in the spinal cord, and the brain and the CNS affect the immune system. The vagus nerve plays essential roles such as in the vagal immune reflex and for mediating information from the distal portion of the small intestine to pain circuits in the brain. Activity in the vagus nerve activates the forebrain cholinergic system through its connections to the nucleus of Meynert.

Stress affects many bodily functions, and recent studies have revealed extensive effects on different systems from the gut to many features of the CNS.

Neurotransmitters involved in emotions and mood diseases

Each of the three neurotransmitters, dopamine, norepinephrine, and serotonin, has widely different functions, and the effect of these three important neurotransmitters overlaps in many aspects. Many of these functions are directly or indirectly related to emotions and the diseases of mood (Figure 3.36).

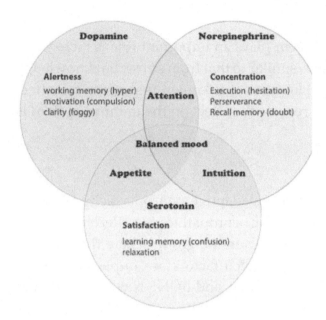

Figure 3.36 Three neurotransmitters and their effects on attention, alertness, and satisfaction. (Artwork by Liliana Cabrera.)

The inflammatory reflex

The inflammatory reflex is complex and involves many structures in the brain such as area postrema (AP); the dorsal motor nucleus of the vagus nerve (DMN); nucleus ambiguus (NA); nucleus of the solitary tract (NST), and enzymes such as acetylcholinesterase (AChE). It was described by Pavlov et al. 2012 [151].

Neuronal interconnections between the nucleus of the solitary tract (NST), area postrema (AP), the dorsal motor nucleus of the vagus nerve (DMN), nucleus ambiguus (NA), and higher forebrain regions integrate afferent signaling and efferent vagus nerve-mediated immunoregulatory output.

Efferent vagus nerve cholinergic output to the spleen, liver, and gastrointestinal tract (blue) regulates immune activation and suppresses proinflammatory cytokine release (dotted red lines). This efferent cholinergic arm of the inflammatory reflex can be activated in the brain through mAChR-mediated mechanisms triggered by muscarinic acetylcholine receptor (mAChR) ligands and acetylcholinesterase (AChE) inhibitors, such as galantamine.

Subcortical structures can modulate cortical activity

Subcortical structures that are involved in the non-conscious perception of emotions can modulate cortical activity either directly or indirectly. The amygdala has direct connections to cortical visual areas in the ventral stream, to the orbitofrontal and anterior cingulate cortices, which are involved in the conscious perception of emotions, and to the frontoparietal network, which is involved in attention.

Pain can modulate emotions

Emotion and cognition are influenced by pain, and pain is influenced by emotion and cognition. These influences are both facilitating and suppressive, which opens many possibilities of interactions. A negative emotional state can lead to increased pain, whereas a positive state can reduce pain. Similarly, cognitive states such as attention and memory can either increase or decrease pain. Of course, emotions and cognition can interact reciprocally as well [152].

The illustration, Figure 3.37, depicts how emotion can both decrease and increases pain, and that pain can decrease emotions. Cognition can decrease and increase pain, and pain can decrease cognition

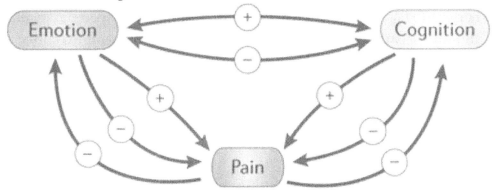

Figure 3.37 Illustrations of an example of feedback loops between pain, emotions, and cognition. The minus sign refers to an adverse effect, and the plus sign refers to a positive effect. (Adapted from: Bushnell, Ceko, and Low, Bushnell, *Cognitive and emotional control of pain and its disruption in chronic pain.* Nature Reviews Neuroscience, 2013.**)** [152].

Also, cognition can modulate (increase and decrease) emotions and vice versa.

The maternal brain and reproduction

One of the aims of ongoing research in maternal affective neuroscience is to examine complex models of maternal psychopathology and parenting. This will aid in identifying neural, behavioral, and psychosocial predictors of postpartum affective disorders and maternal caregiving behaviors. Application of such predictors to mothers before pregnancy or perinatally shows promise concerning the design of treatments. These predictors also show promise regarding early interventions to improve outcomes for mothers and children.

Plasticity in the maternal brain

Early life factors, such as the experience of parental warmth and previous mental health, may affect many neural circuits in the brain. It is possible to alter the function of some of these circuits by the activation of neuroplasticity. These circuits can regulate the function of brain circuits that control maternal affective capacity and caregiving outcomes. Some of these adaptable circuits are overlapping and include circuits that control emotion and the processing of the response to emotional stimuli.

The activation of neuroplasticity in the parts of the brain that regulates maternal mental capacity and caregiving involve many neural circuits, including the nuclei of the amygdala, the dorsal anterior cingulate cortex (dACC), the ventral (vACC) and the prefrontal cortex (PFC). Extensive processing occurs in the amygdala, insula, ventral striatum, and other structures that are related to the cortical executive function (dorsolateral prefrontal cortex, DLPFC).

Empathy is associated with the medial prefrontal cortex (MPFC), precuneus, and superior temporal sulcus circuits. The circuits include the amygdala, the dorsal anterior cingulate cortex (ACC), the ventral ACC, and the prefrontal cortex (PFC). The processing is done in the amygdala, insula, and ventral striatum (VS) working with cortical executive function involving the ventrolateral prefrontal cortex (VLPFC) and the dorsolateral prefrontal cortex (DLPFC).

A putative model of relationships between multiple early-life factors is shown in Figure 3.38. The relationships are shown with the representative brain circuits and maternal and caregiving outcomes. The following network-brain region groupings are emphasized: executive control (DLPFC), ventrolateral prefrontal cortex (VLPFC), fear regulation, dorsal anterior cingulate cortex (ACC), ventral anterior cingulate cortex, hippocampus, and salience/fear processing amygdala, and insula.

Note: some related circuitries are not shown and that other biological mechanisms, specifically genetic, are also left out for the sake of simplicity. The effect of oxytocin on the amygdala is also omitted, but that is discussed below. Brain circuits involved in the regulation of maternal caregiving are shown in Figure 3.38.

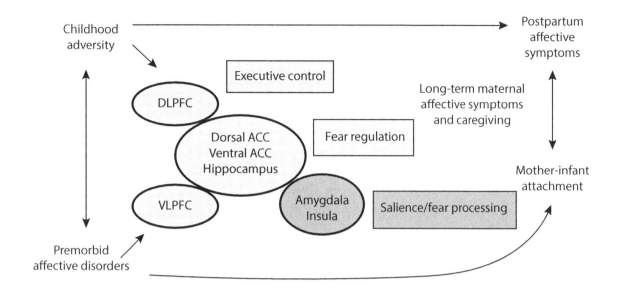

Figure 3.38 A neural model of emotion regulation illustrating possible functional and structural abnormalities in neural systems implicated in the voluntary and automatic subprocesses of emotion regulation in adult bipolar disorder. (Based on: Phillips, et. al 2008, [153].) (Artwork by Liliana Cabrera.)

The left-sided ventromedial prefrontal cortical (VMPFC) regions are the predominate location of the abnormalities in neural systems that are related to the regulation of emotions. These brain regions may be implicated in the regulation of emotion that may cause some forms of mood instability such as that see in adult individuals with bipolar disorders (BD). Persons with such symptoms may also have changes in the gray matter in the left orbital frontal cortex (OFC) and the number of white matter fibers connecting the left OFC may also be affected. The dorsolateral prefrontal cortex (DLPFC), the dorsomedial prefrontal cortex (MdPFC), the anterior cingulate gyrus (ACG), ventrolateral prefrontal cortex (VLPFC), the OFC and the hippocampus-parahippocampus region are also involved [153].

Oxytocin, vasopressin (ADH), dopamine and melatonin

Oxytocin has many functions that are related to reproduction. For example, oxytocin regulates peripheral functions such as the milk letdown reflex and uterine contractions in females as well as vasoconstriction and water retention in both males and females. Oxytocin is perhaps best known for its use to induce uterus contractions in a pregnant woman.

Oxytocin and vasopressin are released from the pituitary gland into the blood stream. Oxytocin and vasopressin are also released throughout the brain to regulate a variety of complex social behaviors, including social recognition, mating, bonding, parenting, and social buffering [154]. Thus, it has been shown that positive social interactions, such as physical contact, are associated with oxytocin release.

The role of oxytocin in reproduction

Oxytocin, best known for its use to induce uterus contractions in a pregnant woman, has many functions related to reproduction. Oxytocin also influences the function of the amygdala. Specifically, it augments the gain in the amygdala and attenuates sensory precision in the hypothalamus.

Oxytocin reduces amygdala and HPA axis reactivity to social stressors, and as such it is an essential mediator of the anxiolytic and stress-protective effects of positive social interaction or "social buffering".

As a mediator of successful reproduction, oxytocin plays a crucial role in establishing a normal socio-sexual behavior, not only affecting the reproductive system, but also assisting other parts of the brain. Oxytocin mediates a top-down modulation, and it may play a role similar to other modulators of NMDA receptor function. Oxytocin does this not only in the cortex but also in the autonomic nervous system and associated subcortical regions of the brain. Acquisition or learning of hierarchical models may depend upon the oxytocin dependent selection of cues with interoceptive associations or significance in the environment.

Oxytocin and emotion

Several studies have found that oxytocin affects emotions, including fear and anxiety. A recent study showed that oxytocin and GABA have similar effects on maternal anxiety and depressions [155]. The same study, performed in laboratory rats, showed that increased activity at the ventral striatum (VS) receptor contributes to the postpartum suppression of anxiety-related behavior which is mediated by physical contact with one's offspring. Failure or incorrect regulation of GABAergic signaling was found to result in deficits in maternal care and anxiety-like and depression-like behaviors during the postpartum period.

To reiterate, oxytocin has many other effects such as augmenting the gain in the amygdala and attenuating sensory precision in the hypothalamic region (Figure 3.39.) This figure illustrates the manifold of possible actions of oxytocin.

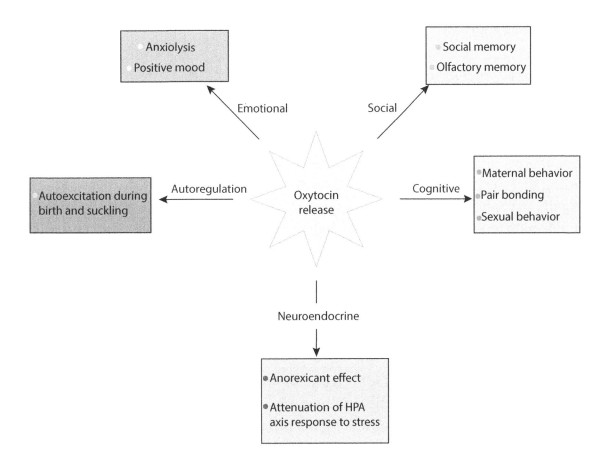

Figure 3.39 Some effects of a release of oxytocin. (Based on: Romano, et al., 2016.) [156] (Artwork by Liliana Cabrera.)

Along with "social buffering", oxytocin influences the establishment of the perception of one's "self", and that is an indication that oxytocin may enable the neuronal plasticity necessary for acquiring the emotional and social "self" [157]. A failure to modulate the precision of interoceptive cues might, therefore, lead to the impaired acquisition of deep hierarchical models that underwrite a sense of self. It may also lead to impaired acquisition of the subsequent sensory attenuation crucial to developing central coherence and negotiating social interactions [158].

Role of oxytocin in reproduction

Oxytocin plays a vital role in creating and forming a person's socio-sexual behavior. Female offspring of monkeys [159] that experience stress and anxiety during early postnatal days tend to display poor maternal care behavior later on in life. Parental care mediates the effects of environmental adversity on the development of the nervous system and, therefore, early postnatal experiences may subdue the genetic predispositions [160] [161].

Oxytocin and learning

Learning represents the updating of associations and connection strengths responsible for generating top-down predictions. The prefrontal cortex, specifically the anterior cingulate and the ventromedial prefrontal region, can then integrate the ensuing representations as part of hierarchical inference that underlies emotional awareness, regarding that the brain works as a prediction machine [123].

Again, oxytocin is a mediator of successful reproduction but also plays a crucial role in establishing a normal socio-sexual behavior, not only affecting the reproductive system, but also in other parts of the brain. Parental care mediates the effects of environmental adversity on the development of the nervous system and, therefore, early postnatal experiences may subdue the genetic predispositions [160] [161]. Also, female offspring of monkeys [159] that experience stress and anxiety during early postnatal days tend to display poor maternal care behavior later on in life.

Similarities and differences between oxytocin and vasopressin (ADH)

Vasopressin (ADH), an antidiuretic hormone, and arginine vasopressin (AVP) are best known for their effects in suppressing the production of urine by the kidneys. ADH is synthesized in the hypothalamus as a peptide prohormone and then converted to AVP. Oxytocin and vasopressin are both released from the pituitary gland into the bloodstream to regulate peripheral functions such as the milk letdown reflex and uterine contractions in females as well as vasoconstriction and water retention in both males and females. Also, oxytocin and vasopressin are also both released throughout the brain to regulate a variety of complex social behaviors including social recognition, mating, bonding, parenting, and social buffering [154].

Vasopressin has a similar effect to that of oxytocin on social buffering that is, and different functions regarding mating, bonding, and parenting [154]. Oxytocin (OT) and arginine vasopressin (AVP) have the same targets in social buffering related to the function of the paraventricular nucleus of the hypothalamus (PVN); OT and AVP have different targets social recognition, mating, bonding, and parenting. The most significant difference between OT and AVP is regarding mating. The difference between the targets and the functions of OT and AVP were studied in the prairie vole [154].

Dopamine

Dopamine is a neurotransmitter that belongs to the catecholamine family. Dopamine is released into the bloodstream from the substantia nigra in the basal ganglia, the ventral tegmental area, and the nucleus accumbens of the reward system. Also, dopamine is released by nerve cells, and it is involved in many basic functions related to emotions such as pleasure, alertness, attention that motivates, motor function desires, and rewards including addiction. For dopamine, the same complementary effects are illustrated in the dorsal and ventral striatum, mediated by D1 (go pathway) and D2 (no-go pathway) receptors, respectively.

Dopamine is critical for many processes that drive learning and memory, including motivation, prediction error, incentive salience, memory consolidation, and response output. Dopamine is perhaps best known for its effect on the motor system illustrated in the symptoms of Parkinson's disease. The dopaminergic diffuse modulatory systems are arising from substantia nigra and the ventral tegmental area. Additionally, dopamine is involved with addictions of various kinds through its effect on the reward system of the brain.

Deficits in oxytocin and dopamine

It has been suggested that deficits of oxytocin lead to a failure of interoceptive processing, while deficits of dopamine, in Parkinson's disease for example, compromise proprioceptive processing and the initiation of action [162].

Glutamatic, Dopaminergic, and GABAergic connections

The primary reward circuit includes dopaminergic projections from the VTA to the NAc, which release dopamine in response to reward-related stimuli and aversion-related stimuli in some cases. There are also GABAergic projections from the NAc to the VTA. These projections go through the direct pathway (mediated by D1-type medium spiny neurons (MSNs)) and directly innervate the VTA, whereas projections through the indirect pathway (mediated by D2-type MSNs) innervate the VTA via intervening GABAergic neurons in the ventral pallidum (not shown).

These monosynaptic circuits for NAc and VTA are shown in Figure 3.40. Figure 3.40 also shows RTMg, rostromedial tegmentum [163], and the glutamatic, dopaminergic, and GABAergic connections in the rat.

Significant dopaminergic, glutamatergic and GABAergic connections to and from the ventral tegmental area (VTA) and nucleus accumbens (NAc) in the rodent brain [163].

Nucleus accumbens NAc receives glutamatic input from:
>   Medial prefrontal cortex
>   Hippocampus
>   Amygdala

Dopaminergic input from:
>   The ventral tegmental area

Nucleus accumbens sends:

GABAergic signals to:
the VTA and LH

The ventral tegmental area (VTA) receives glutamatic input from:
>   LHb
>   LH and

GABAergic input from:
>   NAc

The (VTA) sends:

Dopaminergic signals to:
>   mPFC,
>   Hippocampus,
>   NAc and amygdala

The various glutamatergic inputs control aspects of reward-related perception and memory. The glutamatergic circuit from the LH to the VTA is also mediated by orexin (not shown). The NAc receives dense innervation from glutamatergic monosynaptic circuits from the medial prefrontal cortex (mPFC), hippocampus (Hipp), amygdala (Amy), and other regions as well. Also, the NAc contains numerous types of interneurons. The VTA receives such inputs from the lateral dorsal tegmentum (LDTg), lateral habenula (LHb), and lateral hypothalamus (LH), as well as both GABAergic and glutamatergic connections from the extended amygdala.

The effect of estrogens

Lynch and colleagues [164] studied ovariectomized adult female rats and found that estradiol increases fear generalization through an effect on fear memory retrieval mechanisms by the activation of ERβ [164]. The use and administration estrogen for the treatment of affective disorders has been discouraged because of the belief that estrogen increases the risk of breast cancer [165]. Recent studies, however, have indicated that these concerns may have been overestimated.

Animal studies have shown that estradiol affect the medial prefrontal cortex (mPFC) and the function of a part of the hippocampus (dorsal cornu ammonis 1) (dCA1). The ventral CA1 influences the basolateral amygdala (BLA) which in turn has reciprocal connections with mPFC which receives input the dCA1 [164].) This means that estradiol can affect the dorsal CA cells in the hippocampus and from there affects the basolateral nucleus of the amygdala.

Estradiol also affects the medial prefrontal cortex (mPFC) directly. Additionally, the mPFC influences the BLA. Estrogens increase fear generalization through the activation of the estrogen receptor-beta along with a genomic effect on fear memory retrieval, both within the dorsal hippocampus and anterior cingulate cortex. It has been found that fluctuating estradiol states, one's vulnerability to fear, and anxiety disorders are correlated. It is, however, not known where, how, and when estradiol modulates the fear extinction network. Investigating these questions may provide new options for targeted areas of treatment. Thus, more effective treatment and therapy in the clinic.

Studies have shown evidence that estrogens affect emotional memory [166]. This means that estrogen can also affect emotions, including fear. Also, estrogen levels can regulate HPA responses. As a result, HPA activity can inhibit estrogen secretion. In the HPG axis, the hypothalamus produces the gonadotropin-releasing hormone, which binds to receptors within the anterior pituitary to release luteinizing hormone (LH) and follicle-stimulating hormone (FSH). LH and FSH then stimulate the gonads for the release of estrogen and progesterone, as well as a small amount of testosterone, in females. These systems are also indirectly connected to the hippocampus, mPFC, and amygdala, which are critical regions in fear circuitry that are affected by stress and estrogen.

Women are more vulnerable than men to stress-based and fear-based disorders, such as anxiety and post-traumatic stress disorder. That may indicate that female reproductive hormones may play a role in those disorders. However, the biology of sex differences in emotions, such as fear, remains unclear. The neurobiological mechanisms of fear and stress in learning and memory processes have been extensively studied, and the crosstalk between these systems is beginning to explain the disproportionate rates of incidence and the differences in symptomatology and remission within these psychopathologies [146].

Estrogen appears to modulate activity within the vmPFC and interleukin (IL) to enhance fear extinction memory. The rodent prelimbic (PL) and infralimbic (IL) areas of the medial prefrontal cortex (mPFC) appear to be homologous with the human dorsal anterior cingulate cortex (dACC) and ventromedial prefrontal cortex (vmPFC), respectively.

Future directions for exploring the role of estradiol in fear extinction and psychopathology

An apparent correlation between fluctuating estradiol states and vulnerability for fear and anxiety disorders necessitates further research into where, how, and when estradiol modulates the fear extinction network (Figure 3.40). Investigating these questions may provide new options for targeted, and thus more effective, treatment and therapy in the clinic.

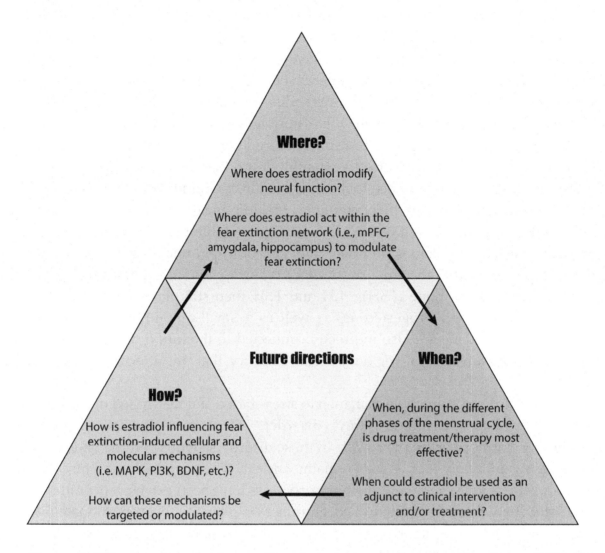

Figure 3.40 Suggestions for future directions for exploring the role of estradiol in fear extinction and psychopathology. BDNF, brain-derived neurotrophic factor; MAPK, mitogen-activated protein kinase; mPFC, medial prefrontal cortex; PI3K, phosphoinositide 3-kinase. (Based on: Cover et al, 2014.) [167]. (Artwork by Liliana Cabrera.)

Melatonin

Melatonin is a hormone secreted primarily by the pineal gland, but other organs also contribute such as the gut, retina, and leukocytes in the production of melatonin. Melatonin is involved in circadian rhythms. Additionally, melatonin is modulated by the amount of light and secretion increases when it is dark. It does not induce sleep, but it signals when it is time to sleep. Melatonin is involved in sleep disturbances, mainly because of light suppressing it [168].

Melatonin is also an antioxidant. Melatonin boosts the immune system, and it has been shown to decrease the risk of some cancers such as breast cancer [169]. Melatonin also has some effects on mood disturbances. Melatonin has a beneficial effect on memory deficits and possibly other symptoms in Alzheimer's disease and other dementias [170].

In some people, especially elderly persons, the pineal gland may not secrete adequate amounts of melatonin. Thus, people take melatonin as a supplement which may help them to sleep.

BDNF and cytokines

BDNF levels are known to increase from physical exercise, and it is also known that the specific fatty acids in one's diet can increase BDNF. Especially Omega 3 fatty acid, Alpha-linolenic acid (ALA), a plant-based essential Omega-3, and polyunsaturated fatty acid have been shown to increase BDNF. Due to these fatty acids releasing BDNF, they have anti-depressive effects.

Proinflammatory cytokines can act on many visceral organs of the abdomen. Inflammatory mediators, such as cytokines, are released by activated macrophages and other immune cells when TLRs and NLRs are activated upon immune challenge. These mediators are detected by sensory components of the afferent arm of the inflammatory reflex (red). Neuronal interconnections between the NST, AP, DMN, NA, and higher forebrain regions (not shown) integrate afferent signaling and efferent vagus nerve-mediated immunoregulatory output.

Efferent vagus nerve cholinergic output to the spleen, liver, and gastrointestinal tract (blue) regulates immune activation and suppresses proinflammatory cytokine release (dotted red lines). This efferent cholinergic arm of the inflammatory reflex can be activated in the brain through mAChR-mediated mechanisms which are triggered by mAChR ligands and AChE inhibitors, such as galantamine.

Mood disorders

The amygdala and depression

Pathological changes in the function of the structures of the amygdala are assumed to be essential for causing the symptoms and signs of mood disorders such as depression. Figure 3.41 shows critical connections of the amygdala that may be involved in creating the symptoms of depression. Figure 3.41 also shows that the regions are grouped into four main compartments which reflect the overall behavioral dimensions of major depressive disorder (MDD) and regional targets of various antidepressant treatments. Areas within a compartment all have strong anatomical connections to one another.

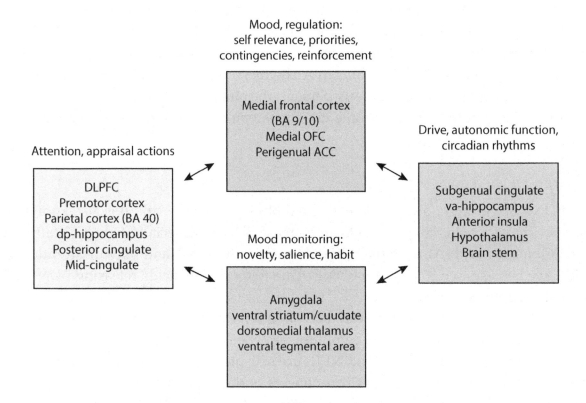

Mood, regulation:
self relevance, priorities,
contingencies, reinforcement

Attention, appraisal actions

Medial frontal cortex
(BA 9/10)
Medial OFC
Perigenual ACC

Drive, autonomic function,
circadian rhythms

DLPFC
Premotor cortex
Parietal cortex (BA 40)
dp-hippocampus
Posterior cingulate
Mid-cingulate

Subgenual cingulate
va-hippocampus
Anterior insula
Hypothalamus
Brain stem

Mood monitoring:
novelty, salience, habit

Amygdala
ventral striatum/cuudate
dorsomedial thalamus
ventral tegmental area

Figure 3.41 The neural architecture of emotion regulation. Mayberg's network model of depression. DLPFC, dorsolateral prefrontal cortex; Connections that are supported by the results of different kinds of studies were summarized by Mayberg. (Based on: Gross and Buck, 2009.) [171, 172]. (Artwork by Liliana Cabrera.)

Disrupted responses of limbic regions, particularly the amygdala, have been associated with biases observed toward negative information in persons with depression, and the ventral ACC and other ventromedial regions represent an important neural interface between cognitive and emotional processing.

Disrupted cortico-limbic circuits, with a critical modulating function for the ventral ACC, may explain both emotional biases and cognitive deficits in depression. This idea has been suggested in an influential model initially proposed by Mayberg. This model has subsequently been developed and shown to have value in both the diagnosis and prediction of treatment response (see Figure 3.41).

Several studies have explored the functional coupling of the amygdala and prefrontal regions. Matthews, for example, reported reduced functional coupling of the amygdala and the supragenual anterior cingulate cortex (ACC) during emotion processing which increased with the severity of depression [173].

Using the technique of structural equation modeling and dynamic causal modeling (DCM) to explore changes in a prespecified network of regions showed abnormal connectivity associated with sad face processing in remitted depression. It also showed disrupted OFC–amygdala connectivity in response to happy faces and a nonsignificant trend toward the same effect of sad faces.

Other investigators found a reduced orbital frontal cortex (OFC) connectivity during emotional processing in never-medicated patients. These investigators suggested that connectivity analysis exploring functional disruption within emotional networks may provide a crucial tool for exploring the affective-cognitive interface in depression [171].

Shared neurobiological foundation in MDD, FM, and NeP

Mood disorders, fibromyalgia (FM), and neuropathic pain (NeP) may have shared systemic consequences [174].) Medication, such as selective serotonin reuptake inhibitors (SSRI), has failed to cause adequate relief from many common forms of depression. Physical exercise is effective in treating the ordinary form of depression and has also shown some beneficial effects on fibromyalgia.

Many studies have confirmed that all three conditions are either precipitated or aggravated by stress (see [174]). FM and NeP are characterized by altered limbic and cortical function in addition to peripheral and central sensitization. Studies based on fMRI results have demonstrated that brain regions, such as the dorsal anterior cingulate, that are central to the experience of negative effects in response to physical pain mediate the distress response in response to the "pain" of social exclusion.

Thus, these findings suggest that emotional and physical pain co-occur often because they share the same central nervous system pathways. Similar functional and structural changes in the amygdala and the hippocampus have been described in MDD, FM, and NeP. Dysfunction of these limbic formations is believed to contribute to the changes in neuroendocrine, autonomic, and immune functioning that may contribute to the generation or worsening of mood and pain symptoms.

Evidence suggests that MDD and FM/NeP mutually amplify each other, and therefore contributing significantly to treatment resistance in both depressive and pain disorders. Timely treatment of MDD may, thus, increase the likelihood of remission and decrease the chances of permanent changes. Also, a full and sustained remission may decrease the chance of recurrences. Unfortunately, the comorbidity of MDD and pain may be an obstacle for early and appropriate diagnosis of MDD. This is supported by the results of a study that found that the resolution of painful symptoms doubled the remission rate in depressed patients. The study showed 36.2% of the participants in those studies who had larger than 50% reduction of pain on a visual analog scale (VAS) attained remission vs. only 17.8% of the individuals who had less than 50% reduction on the VAS [174].

Major depressive disorders, fibromyalgia, and chronic neuropathic pain may have shared systemic consequences.

Major depressive disorders (MDD), fibromyalgia (FM), and neuropathic pain (NeP) may have compromised the homeostatic function of prefrontal cortical-limbic circuitry. This feature appears to disrupt autonomic, neuroendocrine, and neuroimmune regulation. Stress, pain, and depression lead to an excessive and untimely release of a corticotropin-releasing hormone (CRH), an adrenocorticotropic hormone (ACTH), and glucocorticoids. That release of hormones and glucocorticoids contributes to immune activation and the release of proinflammatory cytokines such as TNF-alpha, IL-1, and IL-6 contribute to that (Figure 3.42).

There is also evidence from studies that show that excessive sympathetic activation likely plays a role in the etiology of all three of these diseases (MDD, FM, and NeP). All three conditions are associated with the disturbed neuron-glia relationship, glutamatergic dysregulation, and alterations in intracellular signaling cascades and neurotrophic trafficking (Figure 3.42). Both depression and anxiety are prevalent in FM and NeP. Also, depression and anxiety may intensify the experience of pain.

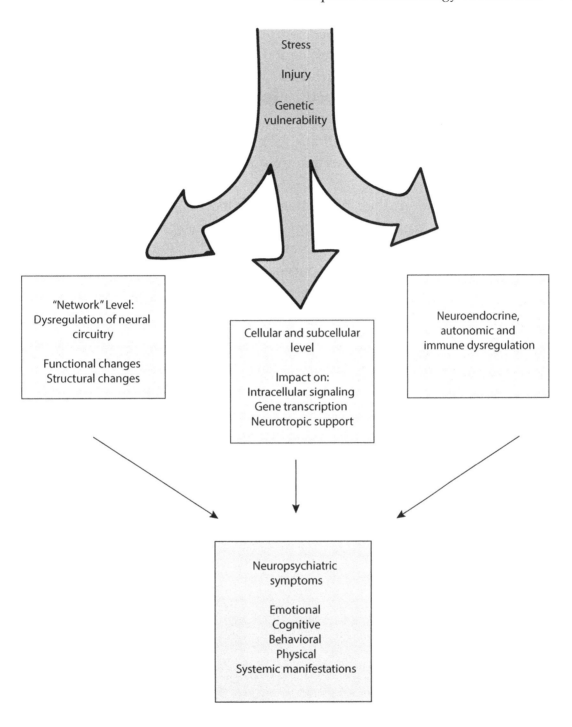

Figure 3.42 An integrated view of the shared neurobiological underpinnings of major depression diseases (MDD), fibromyalgia (FM) and neuropathic pain (NeP) [174]. (Based on: Maletic and Raison, 2009.) [174] (Artwork by Liliana Cabrera.)

Sympathetic over-activity, combined with diminished parasympathetic tone, leads to many functional changes in combination with the downregulation of central glucocorticoid receptor sensitivity which leads to further disruption of feedback control of the hypothalamic-pituitary-adrenal (HPA) axis and the immune system. Disturbances of serotonin (5HT), norepinephrine (NE), and dopamine (DA) transmission may also occur, thus impairing regulatory feedback loops that turn off the stress response (see [174].) Inflammatory cytokines further interfere with monoaminergic and neurotrophic signaling.

Elevated mediators of the inflammatory response, combined with excessive sympathetic tone, may further impact the dorsal column processing pain signals by contributing to the activation of microglia and astroglia. Activated microglia exchange signals with astrocytes and nociceptive neurons. This exchange amplifies the pain-related transmission of glutamate (Glu), substance P (SP), adenosine triphosphate (ATP), brain-derived neurotrophic factor (BDNF), pro-inflammatory cytokines (IL-1, IL-6, IL-8, TNF-alpha, nitrogen oxide (NO), and prostaglandins (PGs).

A compromised homeostatic function of prefrontal cortical-limbic circuitry in MDD, FM, and NeP appears to disrupt autonomic, neuroendocrine, and neuroimmune regulation (Figure 3.43).

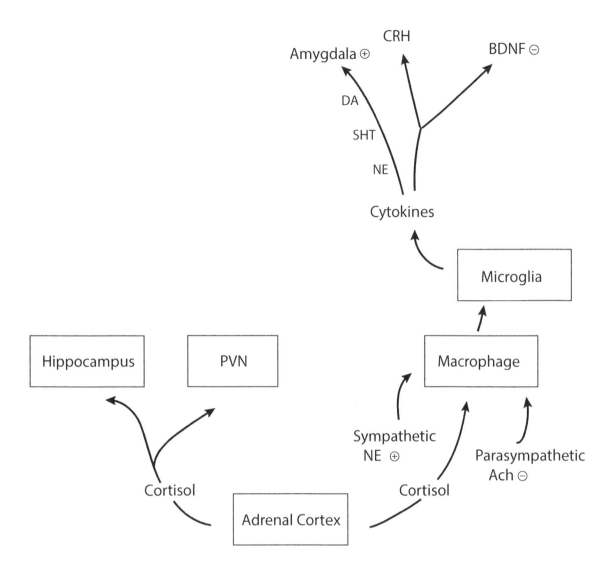

Figure 3.43 A simplified description of interactions between neural transmitters that are involved in various kinds of neural disorder where the immune system plays important roles. (Based on Maletic, and Raison, 2009 [174].) (Artwork by Liliana Cabrera.)

The high rates of comorbidity observed between major depression, fibromyalgia, and neuropathic pain may be the result of the fact that these disorders share multiple biological and environmental underpinnings suggesting that they result from similar genetic vulnerabilities in vulnerable persons. The development of comorbidity may be promoted by psychosocial stress and illness, promoting, relative resistance to the actions of glucocorticoids, increased sympathetic/decreased parasympathetic activity, and increased production and release of proinflammatory mediators.

Dysregulation of stress/inflammatory pathways that promotes alterations in brain circuitry may also contribute to the modulation of mood, pain, and the stress response. These changes, in turn, have been associated with the related processes of central sensitization in pain disorders and "kindling" in depression, both of which may account for the progressive and self-perpetuating nature of these disorders, especially when inadequately treated [174].

Affective cognition and its disruption in mood disorders

Although the amygdala is a central structure in emotional processing, the ACC and OFC are also critical substrates of the affective-cognitive interface. It is also evident that the exact demands of the tasks are essential factors in determining the extent to which these structures, and other structures, are engaged. Furthermore, recent studies emphasize the importance of connectivity [82]. It is not just responses of the amygdala, ACC, or OFC that are important, but rather the strength of connectivity between these regions, and their modulatory control of other regions. For example, modulation of the amygdala-hippocampal connections seems to be a critical substrate of emotional memory, whereas connectivity between the amygdala and the visual cortex is critical in emotional face expression (see Figure 3.44).

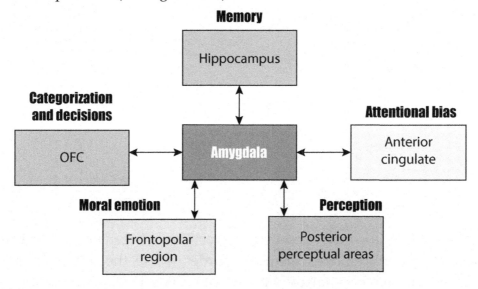

Figure 3.44 Key connections of the amygdala that may be differentially probed by affective cognition paradigms. The figure is a schematic only and does not purport to display all amygdala connections or the interconnection between any other components[171]. (Based on: Elliott, et al, 2011.) [171] (Artwork by Liliana Cabrera.)

Protective mechanisms of the body

Specific circuits in the brain aid in the detection of threats of various kinds, and other circuits aid in the survival response after threats. A fast and direct action, through the non-classical or "fast and dirty" system, enables a fast and effective execution of signals from the environment and the body. There are also possibilities for a more refined selection of responses to various forms of fear through the classical or "slow and accurate" system.

There are many mechanisms in the nervous system that are specifically devoted to protecting vital functions in the case of harm to the body or an instance when there are risks of various kinds of dangers. Additionally, emotions can trigger the body's protective mechanisms. Pain is a potent initiator of these protective mechanisms, and pain response has been studied in more detail than responses triggered by fear [175]. Warning about the dangers of various kinds come in the form of pain, illness response, hunger, and thirst. Fear plays a major role as a warning signal for hazards that may be a threat to the body in the future.

Bodily reactions to fearful situations, such as the "freezing" or paralysis reaction, occurs in response to a perceived danger may be more difficult to explain. Freezing is an old protective action that serves the purpose of avoiding harm because it may be safer in a dangerous condition to not move. Not moving may be beneficial because it can decrease the risk of an attack, but it also means that a person cannot flee, which may sometimes be a disadvantage. The "freezing" reaction is more pronounced in some animals, such as the rat, which can easily freeze but it does indeed exist in people as well. In our modern societies with modern threats, it may no longer be beneficial not to move when a person is threatened because it may prevent a person from escaping a dangerous situation. Increased heart rate and increased blood pressure are common reactions to stress. Also, some people faint in response to fearful situations. These reactions, freezing and particular fainting, seem to be the opposite of what would be expected in a situation of danger to a person.

The opposite reaction of "freezing" would be for one to engage in a "fight" response. For fighting, fear puts the body in a condition that is beneficial to fight by redirecting nutrients and oxygen to various muscles from the skin and internal organs except for the heart and the brain.

Fighting or the urge to run away are examples of other automatic reactions to fear stimuli. The fight or flight reaction, that has typical responses from the autonomic nervous system, puts the body into an alarm mode with increased blood pressure, increased heart rate, a diversion of blood from the skin to avoid bleeding if injured, and a diversion of blood from the digestive organs to save energy for muscles and the brain.

These are all reactions that occur in response to increased activation of the sympathetic nervous systems and decreased activity of the parasympathetic nervous system. Extreme fear triggers these conditions. These conditions serve to protect a person, or an animal, from harm and to put the person in a better condition for fighting.

Fear and anxiety can activate the autonomic nervous system through increasing the sympathetic activity and decreasing the parasympathetic activity, but it can also decrease the effectiveness of the immune system. Fear also greatly affects a person's mood. Unfortunately, it often reduces a person's quality of life. Fear plays a part in the body's alarm systems, and it can detect immediate dangers and warn about future dangers. Additionally, it can initiate actions to protect the body from the damage of various kinds.

Activation of central circuits leads to endocrine, autonomic, somatosensory, and illness responses as well as pain and hyperalgesia effects as a component of the illness response. The afferent feedback from the body tissue is neural, hormonal, and mediated by cytokines from the immune system. The central circuits are adapted, by the forebrain, to specific environmental situations.

When a survival circuit is triggered, specific circuits in the brain are activated and many consequences may result. Also, innate behavioral responses are potentially activated as well as the autonomic nervous system (ANS) responses and hormonal responses. Each one of these generates feedback to the brain. After neuromodulator systems are activated, they begin to regulate excitability and neurotransmission throughout the brain. Thus, the motivation system initiates goal-directed instrumental behavior (Figure 3.45). Sensory, cognitive, and explicit memory systems are also affected, which lead to an enhanced level of attention to relevant stimuli and the formation of new explicit memories (memories formed by the hippocampus and related cortical areas) and implicit memories (memories formed within the survival circuit) [102].

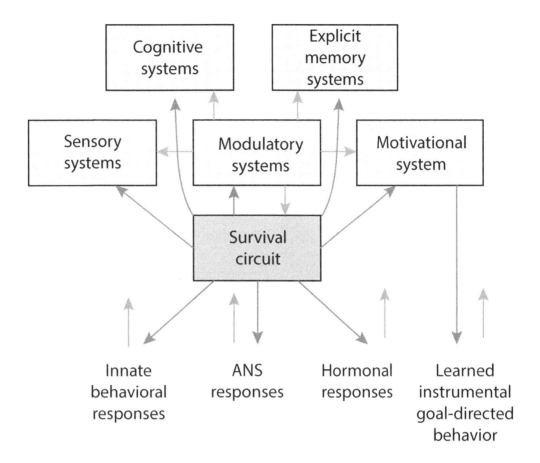

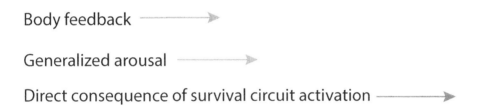

Figure 3.45 Consequences of survival circuit activation. (Based on: LeDoux, 2012.) [102]. (Artwork by Liliana Cabrera.)

When a survival circuit trigger activates a survival circuit, many consequences follow. (1) Innate behavioral responses may be activated, together with activation of parts of the autonomic nervous system (ANS) and specific hormonal responses. These reactions are elicited from the brain and information about their activation generate feedback to the brain. (2) Neuromodulator systems are activated and begin to regulate the excitability of specific circuits in the brain, and the neurotransmission in certain circuits in the brain is also regulated. (3) Goal-directed instrumental behavior is initiated by the motivation system.

Neural circuits that serve to protect the body

There are several systems in our brain that serve the purpose of aiding in survival responses and actions one takes when a threat of one kind or another occurs. In humans, that response to various forms of fear occurs through a fast-direct action via the non-classical or "fast and dirty" system, or through a more refined action via the classical or "slow and accurate" system.

Activations of the body's protective mechanisms trigger the central representation of protective bodily functions which in turn can control many functions as illustrated in Figure 3.46 such as various autonomic functions, endocrine functions, and motor functions.

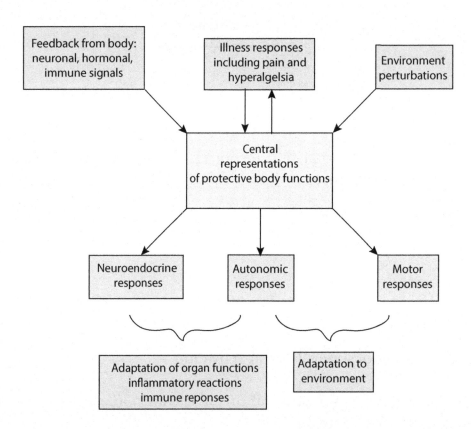

Figure 3.46 Activation of central circuits leads to endocrine, autonomic, somatosensory and illness responses as well as to pain and hyperalgesia (as a component of the illness response). (Based on: Jänig, and Häbler, 2000.) [175]. (Artwork by Liliana Cabrera.)

Neural, hormonal, and immune signals play a major role in modulating the response from the body. Also, environmental functions can modulate the response from the body's protective system. The illness responses and pain signals that activate the body's protective mechanisms also receive input from the body that may strengthen or weakened the activation of the protective functions.

Which hemisphere of the brain is involved in emotions?

The two sides of the brain, the left and right hemisphere, communicate with one another via the corpus callosum. The two hemispheres have slightly different functions for many everyday tasks. Fear is not an exception. Dr. Noesselt and his team studied the differences in activation of structures in the left and right hemisphere in response to fearful versus neutral faces by presenting the participants in the study with faces bilaterally and orthogonally, thus manipulating whether each hemifield showed a fearful or neutral expression before the presentation of a checkerboard target [176].

The results from behavioral studies and fMRI studies showed evidence of right-lateralized emotional processing during bilateral stimulation. Connectivity studies using fMRI imaging showed signs of enhanced coupling of the amygdala to the right-hemispheric extrastriate cortex. Other studies have found distinct differences between the processing of speech. These studies have shown that the left brain is more engaged in logical and fact-based thoughts such as science and mathematics, whereas the right brain is more involved in emotional issues and creativity (Figure 3.47.)

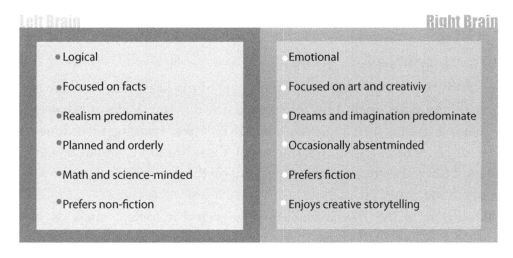

Figure 3.47 Differences between the left and right brain functions.

The left brain uses logic, detail-oriented thinking, facts, words and language, math and science, comprehension of knowledge, acknowledgements, object name recognition, and reality-based thoughts. The right brain uses feelings, big picture-oriented ideas, imagination, symbols and images, intuition, appreciation, spatial perception, object function knowledge, fantasy-based thoughts, conceptual possibilities, and risk-taking ideas.

Recent developments in neuroscience

Many more people are now engaged in neuroscience research than just a few years ago. New technology has made studies and experiments, that were not possible a few years ago, possible and are being used in many laboratories. People's curiosity and ingenuity have been the driving force of these advancements, and serendipitous observations made in the research laboratories and during treatments of different kinds of diseases have brought surprising results.

In the past 30 years, we have witnessed an explosion of research in affective neuroscience that has addressed questions such as: Which brain systems underlie emotions? How do differences in these systems relate to differences in the emotional experience of individuals? Do different regions underlie different emotions, or are all emotions a function of the same underlying brain circuitry? How does emotion processing in the brain relate to bodily changes associated with emotion? Moreover, how does emotion processing in the brain interact with cognition, motor behavior, language, and motivation? [82].

Anatomical location of neural activity

Most methods that aim at studying which part of the brain that become activated in connection with emotions, including fear, make use of some form of what is known as functional imaging such as functional MRI or fMRI. These imaging techniques use blood oxygen level dependent (BOLD) imaging. All these methods make use of inducing changes in blood flow, and these methods depend on the hypothesis that increased neural activity results in increased blood flow.

In particular, it is the development of new technologies, such as optogenetic techniques, that has provided more details in the exploration of which structures are involved in fear reactions in animals. These techniques make use of light-sensitive proteins to control nerve cells that have been genetically sensitized to light. This technique makes it possible to identify which populations of nerve cells are involved in a specific process such as fear, and this technique also makes it possible to manipulate several populations of nerve cells.

These new techniques have provided much knowledge about the precise neuronal circuits contributing to the neural expression and recovery of conditioned fear behavior. These techniques have made it possible to determine the contributions of distinct brain regions such as the amygdala, prefrontal cortex, hippocampus, and periaqueductal gray matter with greater precision than what was possible with previously created techniques such as fMRI [177].

Functional connectivity

Studies of functional connectivity in the brain have been possible because of the amazing developments of techniques such as magnetoencephalography (MEG) and specialized electroencephalography [37]. Recent studies using these techniques have revealed the importance of functional connections in many parts of the brain [82]. Notably, it has been shown in many studies that functional connections are graded and subject to rapid changes. Such changes in functional conductivity through the activation of neuroplasticity have been related to essential functions such as specific diseases and age-related changes. Many diseases have now been linked to changes in functional connectivity. Studies of the connectivity are now regarded as a new part of the field of neuroscience.

Results of studies reveal the functional connectivity of many parts of the central nervous system are compatible with the results of behavioral studies. For example, it has been possible to identify which one of the main nuclei of the amygdala is involved in fear. In addition, the involvement of the periaqueductal gray (PAG) has been clarified.

The periaqueductal gray or "PAG" is a midbrain structure involved in critical homeostatic neurobiological functions. Also, the PAG plays a central role in pain modulation and is involved in cardio-respiratory control. Using these new imaging techniques, it has been shown that the medial division of the central nucleus of the amygdala projects to the PAG. (For a review of these recent studies see [177].)

It has been pointed out that it is not just that there are connections between different groups of neurons that are important, but rather the strength of connectivity between regions is an essential measure of functionality in the central nervous system. For example, modulation of the amygdala-hippocampal connections seems to be a key indicator of emotional memory, whereas connectivity between the amygdala and the visual cortex is critical in emotion perception in faces [171].

Connections between different structures in the brain are almost always two-way (reciprocal connections). For example, there are connections going from the thalamus to the amygdala, but there are also connections from the amygdala to the thalamus. Two-way connections form loops where information can circulate thus making iterations as a tool for interpretations possible.

It has been seen that the nuclei of the amygdala connect to many different structures in the brain. This is in good agreement with the modern view of the functional organization of the brain because the brain is a distributed system rather than being compartmentalized.

The structures that are primarily associated with fear is the amygdala and the prefrontal cortex, but these structures are also connected to many other structures in the brain. These connections are not always active but can be activated by changes in the efficacy of synapses through the activation of neuroplasticity.

The strength of the connections in the brain is controlled by neuroplasticity and can vary [1]. What happens in the nervous system is, therefore, similar to the electrical wiring in a house. The wiring is always there, but the light only gets started when a light switch is turned on. In many places, the light can be dimmed, and that would correspond to decreasing the strength of the connections in the brain. The wiring in the brain, the nerve fibers, is always there, but only when the switches, or synapses, are turned on does a connection become activated.

The switches are not just on and off, but they can also be gradual as is the case when the light is dimmed. In the brain, it is the changes in the efficacy of the synapses that control the strength of the connections similar to the light dimmers in the electrical wiring in a house controls the light. Activation of neuroplasticity makes these changes in synaptic efficacy.

It is now generally accepted that the brain is a distributed system where various functions involve many different parts of the brain. It has been known for a long time that the brain is plastic and malleable for the most part, but a new understanding of many aspects has emerged recently. For example, it is well-known that activation of neuroplasticity can change many functions, whereas functions such as long-term memory, sexual preference, and handedness seem to be "hard-wired." Many studies have confirmed that maladaptive plasticity plays a vital role in many common diseases such as chronic neuropathic pain, spasticity, and possibly other disorders such as fibromyalgia [134].

The study of connections in the CNS has developed into a subspecialty of neuroscience devoted to "connectivity". Studies of connectivity have revealed that many anatomical connections usually are not functional, and that the strength of functional connections can vary widely. Thus, neuroplasticity makes many functional brain connections dynamic. Changes in connectivity have been related to symptoms and signs of disease such as chronic neuropathic pain, severe tinnitus, and age-related symptoms and signs of diseases. The cholinergic system of the forebrain (the nucleus of Meynert), the vagus nerve, promotes the activation of neuroplasticity. The activation of neuroplasticity through electrical stimulation of the vagus nerve may be used to reverse maladaptive plasticity thus alleviating the symptoms and sign of some common diseases.

Coding of information in the nervous system

Traditionally, it has been assumed that a specific structure is involved in a task. It also was assumed that a specific structure engages the nervous system and that signs of an increase in the neural activity of specific nerve cells can be detected when the task is activated. The earlier focus was on the function of groups of nerve cells, and increased activation has been linked to an increase in the firing rate of nerve cells. Recently, functions have been linked to the synchronization between members of ensembles of nerve cells.

Recent studies have shown that connections between groups of nerve cells are essential indicators of functional activity. This means that the focus has shifted from gray matter to the white matter of the brain.

In general, recent progress in neuroscience has shown that many, or perhaps most, structures of the CNS have many functions. It has also been indicated that a specific function involves many different anatomical structures. Also, a general observation in many recent studies is that many different brain structures are involved in the same specific functions as one another. For example, what structures that were previously regarded as belonging to the motor system are also involved in such tasks as pain. The cerebellum, which was earlier assumed to be devoted to motor functions, has been shown to be involved in cognitive functions as well. It, therefore, does not come as a surprise that the experience of emotions, such as fear, involves many different structures. However, some of these anatomical structures play more critical roles than other structures.

Chapter 4 Neurobiology of fear and anxiety

Neurobiology of fear

The neurobiology of emotions was discussed in Chapter 3. In this chapter, the neurobiology of the brain structures that are specifically involved in the expression of fear and anxiety will be discussed. The functions of these structures and their involvement in the creation of the fear experience are also discussed. This discussion includes a look into anxiety itself and the diseases that are caused by excessive fear.

First, this chapter briefly reviews discoveries regarding the function of some brain structures that are relevant for understanding what happens in the brain when a person experiences fear.

What happens in the brain when a person is afraid, or anxious?

Our knowledge about which brain structures are activated when a person experiences an emotion, such fear or anxiety, is mainly based on studies in animals. Most of these studies were done in rats, but a few studies have involved monkeys. It is, however, not known if animals can feel fear. Yet, in studies of fear where animals are used, the reactions of the animals are common indications of fear.

Many brain systems that are involved in fear and anxiety may be activated, primarily including the different structures of what Joseph LeDoux has called the emotional brain, the insula, and the anterior cingulate cortex (ACC).

Figure 4.1 shows the most important structures that are involved when a person experiences fear and how these structures interact. These structures are similar to those that are involved in emotions as was discussed in the previous chapter.

**Brain systems in fear
LeDoux (1995)**

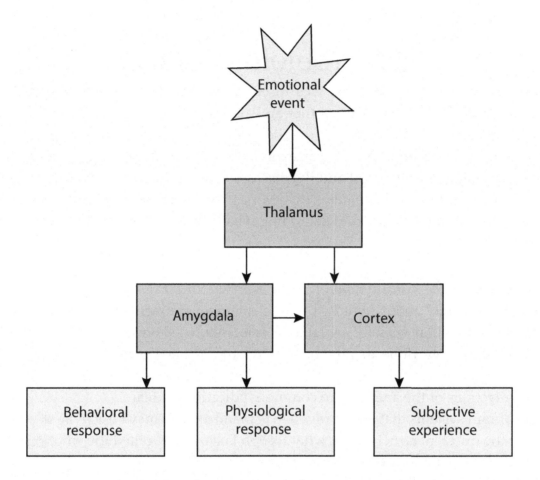

Figure 4.1 The most important structures involved in the creation of the expression of fear and the effect of fear on body functions. (Based on LeDoux, 1995) (Artwork by Liliana Cabrera).

As was discussed in Chapter 3, the activation of structures in the brain in connection with fear has been studied in detail by another well-known neuroscientist, Dr. James McGaugh. Dr. McGaugh studied the memory-related aspects of fear especially [85, 178]. We, therefore, owe these two researchers, Dr. LeDoux and Dr. McGaugh, and their students a recognition for much of our understanding of the changes in the function of the brain that occur in connection with fear.

Functional connections

Functional pathways of fear conditioning (Connectivity)

There are extensive anatomical connections between multiple parts of the brain, but not all connections are functional because there are ineffective synapses [179]. Only relatively recently has it become possible to study the functionality of anatomical connections. While anatomical connections between specific nerve cells either exist or don't exist, there is no in between, functional connections are graded, and these graded connections can various levels of strength [82].

New technology has recently made quantitative studies of the strength of connections in the brain possible [82]. Techniques such as magnetoencephalographic (MEG) recordings, some forms of electroencephalographic (EEG) recordings, and the measurement of very small changes in blood flow, known as functional magnetic resonance imaging, functional MRI, or fMRI, have made it possible to quantitatively determine the strength of connections between large groups of nerve cells in different parts of the brain.

Studies of functional connections have fundamental importance for understanding the organization of the central nervous system. As a result, a new branch of neuroscience has evolved known as the "study of connectivity" in the brain and the spinal cord. Studies of functional connectivity in the brain are possible because of the development of techniques such as MEG recordings.

Studies of connectivity have developed into a quantitative method that is promising for understanding not only the normal functions of the brain, but also for the study of diseases and what causes the symptoms of many diseases. Studies regarding connectivity have revealed that the brain is a distributed system. The extensive functional connections that exist between most regions of the brain are often reciprocal creating loops, where information may circulate. These connections and loops are most likely the basis for the use of iterative methods in interpreting information in the brain. Recent studies of functional connectivity have shown that most tasks, even simple ones, engage many parts of the brain simultaneously. Also, we know that several parts of the brain are involved in most tasks, and some parts of the brain can do more than one task [82].

Recently it has become evident that functional connections in the brain play more critical roles in the many functions of the brain than what was earlier believed. Previously, the function of nerve cells was regarded as responsible for many standard functions of the brain, and abnormal functions were regarded to be the result of a malfunction of nerve cells.

Now, much evidence from many different studies of brain function shows the way in which nerve cells are connected is essential for many normal functions of the brain, and that abnormal brain functions that cause the symptoms of certain diseases can be related to changes in the functional connections between groups of nerve cells. There is considerable evidence that changes in the functional connections in the brain play essential roles in many forms of abnormal reactions [82].

Understanding of how fear that is not related to the real risks of harm is created is essential to our understanding, and it is likely that the fear from harmless matters and fear related to the wrong types of risks can be related to abnormal connections in the brain. It was previously assumed that changes in the function of the nervous system were the result of a change in the function of nerve cells. It has now become evident that changes in the strength of connections are also significant. It is, perhaps, even more important for explaining changes in functions such as those that occur in diseases of various kinds and severity, including aging. Connections in the brain are also necessary for the processing of emotions such as when a person experiences fear.

Some pathways of fear conditioning have been discovered recently, and this is now a hot research topic in neuroscience. If the auditory cortex pathway is lesioned, for example, essential fear conditioning is unaltered, but discrimination is altered. In the discrimination procedure, a sound is paired with a shock and another sound is not paired with a shock. The animals had to rely solely on the thalamus and amygdala for learning, and they could not learn the discrimination. Thus, the two stimuli were indistinguishable. The conclusion is that the cortex is not needed for simple fear conditioning. Instead, the cortex allows us to recognize an object by sight or sound. Thus, allowing us to interpret the environment.

Recent studies indicate that fear, specifically fear which is caused by things that are not predictable, uses a different part of the brain than fear resulting from predictable events. Fear that is caused by events that are only known by their likelihood is mainly associated with the bed nucleus, whereas fear caused by predictable matters engages the amygdala complex mainly. Also, the bed nucleus has ample connections with the award system of the brain.

Functional connections in the brain are altered in fear

Functional connections in the brain are important, and changes in functional connections occur in many situations, including changes that occur during emotions such as fear. It was previously believed that changes in the functions of the brain, such as those that occur during motor and mental activities as well as during diseases, are caused by specific changes in the function of cells in certain parts of the brain. Recent studies seem to indicate that changes in connections between nerve cells are more important than changes in the function of nerve cells.

These indications mean that the focus of pathologies has shifted from a gray matter of the brain, the concentrations of nerve cells or soma, to white matter, axons or nerve fiber tracts. Several studies have addressed the question about functional connections in the brain in connection with fear.

A study by a research group at the University of Alabama showed that conditioned changes in the emotional response to threat, such as aversive unconditioned stimulus, are mediated in part by the prefrontal cortex [15]. This study also showed that the emotional response to a likely threat is less than that of an unpredictable threat which elicits large emotional responses [15].

Knowledge about the connections from cells in the prefrontal cortex to cells in other regions of the brain is essential to understand the neuroscience of fear. Such knowledge helps to understand which neural processes in the brain mediate information about a decrease or increase in the emotional response to a threat.

Neural circuits for context-dependent regulation of fear memory

The amygdala, prefrontal cortex, and hippocampus are brain structures that play extremely crucial roles in fear memory [105, 178]. The hippocampus, which is heavily involved with memory, receives information from the amygdala and then sends information back to the amygdala. Recently it was shown that dynorphin, an opioid that acts through kappa opioid receptors, plays an essential role in the formation and extinction of fear memories in mice and humans [105, 180].

Memories are vital for many of the functions we execute every day. In the same way, memory is crucial for fear. The hippocampus projects directly to the BLA, and this projection may be crucial for the renewal of fear expression in response to an extinguished conditional stimulus. Drugs that reduce the effect from the administration of adrenergic substances, such as adrenaline and noradrenaline, are beta-adrenergic blocking agents.

Studies have shown that the beta-adrenergic blocking agent, propranolol, can block the consolidation and reconsolidation of emotional memories in healthy adults [181]. Propranolol is an old beta-adrenergic blocker that has recently become of interest in the control of fear. Propranolol acts as an anxiolytic. Perhaps this is because it reduces one of several expressions of fear, namely the sympathetic expression.

The effect of propranolol on fear may also be related, at least partly, to its effect of reducing memory consolidation, [130] including the memory of fear related matters [131]. Propanolol, administered together with D-cycloserine, has been shown to enhance behavioral approaches to managing dental anxiety [182]. Many systems can modulate memory consolidation, the autonomic nervous system can modulate memory consolidation, and it can do it by modulating the liberation of adrenergic substances [105].

In a study of connectivity, in connection with fear-potentiated startle (FPS) responses during extinction, Fani and her coworkers showed that extinction of conditioned fear is an associative learning process that involves communication among the hippocampus, medial prefrontal cortex, and amygdala [183]. These investigators used the techniques of diffusion tensor imaging (DTI), probabilistic tractography analyses to examine cingulum, and uncinate fasciculus (UF) structural connectivity in people with psychological trauma exposure. These investigators found a significant negative association between cingulum microstructure and FPS during both early and late extinction. In this study, fear-potentiated startle (FPS) responses during fear conditioning and extinction were assessed via electromyography (EMG) of the right orbicularis oculi muscle.

The primary routes of communication between these areas were found to be specific white matter tracts, the cingulum and uncinate fasciculus (UF). However, no significant correlations were observed between fear-potentiated startle (FPS) and the uncinate fasciculus (UF). Its function is unknown, even though it has been implicated in several psychiatric conditions. The strength of connectivity between the hippocampus and the anterior cingulate cortex (ACC), and also between the amygdala and ventromedial prefrontal cortex (vmPFC), may be influenced by the UF.

The concept that gonadal hormones can modulate the expression of fear and other emotions is known. People with PTSD show deficits in their ability to inhibit conditioned fear responding after extinction training. PTSD is more than twice as prevalent in women compared with men, and there is considerable evidence in the scientific literature describing biological mechanisms underlying these disparities. Additionally, women are 60% more likely to suffer from anxiety disorders than men. The hypothesis may explain, to some extent, why females exhibit increased rates of fear generalization, a process partially driven by estrogens. The fluctuating gonadal hormones in the menstrual cycle, especially estrogen, may play a critical role in fear extinction. Hence, the PTSD vulnerability and symptom severity in women [166]. These authors conclude that estrogen treatment may be a putative pharmacologic adjunct in extinction-based therapies and should be tracked in the menstrual cycle during PTSD treatment.

A summary of the sources of the neural response to fear and the relationship to structures that are believed to be involved in PTSD are shown in Figure 4.2.

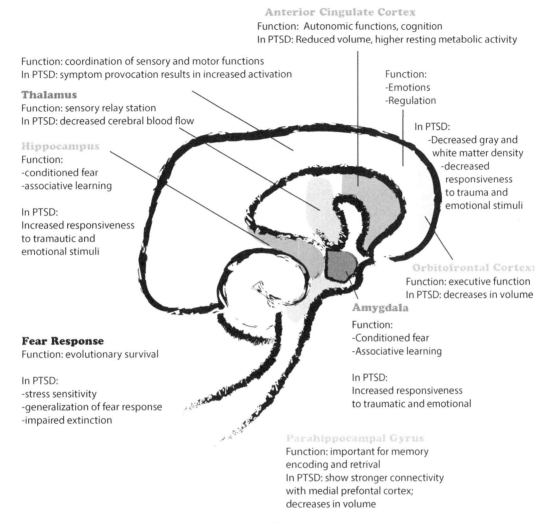

Anterior Cingulate Cortex
Function: Autonomic functions, cognition
In PTSD: Reduced volume, higher resting metabolic activity

Function: coordination of sensory and motor functions
In PTSD: symptom provocation results in increased activation

Function:
-Emotions
-Regulation

Thalamus
Function: sensory relay station
In PTSD: decreased cerebral blood flow

In PTSD:
-Decreased gray and white matter density
-decreased responsiveness to trauma and emotional stimuli

Hippocampus
Function:
-conditioned fear
-associative learning

In PTSD:
Increased responsiveness to tramautic and emotional stimuli

Orbitofrontal Cortex:
Function: executive function
In PTSD: decreases in volume

Amygdala
Function:
-Conditioned fear
-Associative learning

In PTSD:
Increased responsiveness to traumatic and emotional

Fear Response
Function: evolutionary survival

In PTSD:
-stress sensitivity
-generalization of fear response
-impaired extinction

Parahippocampal Gyrus
Function: important for memory encoding and retrival
In PTSD: show stronger connectivity with medial prefontal cortex; decreases in volume

Figure 4.2 A schematic of the human brain illustrating how the limbic system is involved in PTSD (Based on Shin, 2010.) [184]. (Artwork by Liliana Cabrera.)

The prefrontal cortex and the hippocampus both have dense connections to the amygdala, which is important for conditioned fear and associative emotional learning. The PFC is thought to be responsible for reactivating past emotional associations, and the PFC decreases in both responsiveness and density. The hippocampus is thought to play a role in explicit memories regarding traumatic events. The hippocampus also plays a role in mediating learned responses to contextual cues, and in PTSD the hippocampus is decreased in volume and responsiveness to traumatic stimuli. Furthermore, the top-down control of the amygdala by the hippocampus and PFC may result in the increased activation of the amygdala, as is observed in subjects with PTSD. The end result of these neuroanatomical alterations is increased stress sensitivity, generalized fear responses, and impaired extinction.

Other regions, including the anterior cingulate cortex, the orbitofrontal cortex, the parahippocampal gyrus, the thalamus, and the sensorimotor cortex, also play a secondary role in the regulation of fear and PTSD [184].

Long-range excitatory and inhibitory connections between multiple brain areas are involved in generating fear states. The amygdala nuclei that receive sensory input from cortical and thalamic centers are major sites of fear-related neuronal plasticity [185].

Prefrontal cortex control of fear

The role of the prefrontal cortex in fear is twofold: The prelimbic prefrontal cortex is implicated in the generation of the fear response, whereas the infralimbic prefrontal cortex plays a vital role in the extinction of fear learning [53]. The input to these two structures comes from the amygdala and other supplementary structures. These structures gate the expression of fear through their influence on circuits in the amygdala.

Activation of the BLA-to-ventral hippocampus (vHC) pathway is anxiogenic, whereas activation of the BLA-to-central amygdala (CEA) projection is anxiolytic. By contrast, the two parallel ventral BNST (vBNST)-to-ventral tegmental area (VTA) pathways mediate either anxiogenic or anxiolytic behavioral outcomes. Anxiety states are mediated by local and long-range connections between multiple brain areas [185]. However, large parts of the anxiety network remain to be fully characterized regarding the cellular identity and functions as well as the precise local and long-range connectivity using modern circuit-based approaches.

A model of interactions between the different components of the amygdala and the prefrontal cortex and the nucleus accumbens can explain some of what is known about expression and extinction of conditioned fear (Figure 4.3). The model includes known pathways and possible links (marked by asterisks) that collectively account for most of the available evidence [50, 186, 187]. HYP, hypothalamus; LC, locus coeruleus; LS, lateral septum; mPFC, medial prefrontal cortex; ovBNST, oval BNST; PAG, periaqueductal gray; PB, parabrachial nucleus; RN, raphe nuclei.

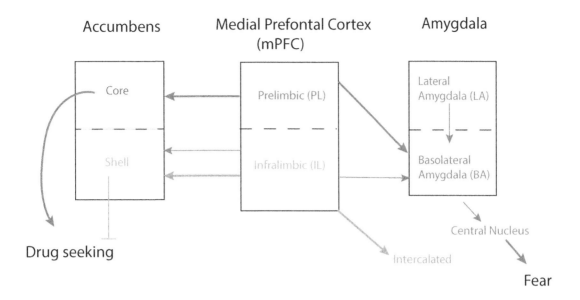

Figure 4.3 A circuit diagram depicting how the medial prefrontal cortex (mPFC) regulates conditioned fear and cocaine-seeking behaviors. The dorsal and ventral subdivisions of the medial prefrontal cortex are shown at the center, with their respective outputs to the amygdala controlling fear shown on the right, and those to the nucleus accumbens controlling cocaine seeking shown on the left [187]. (Based on Kalivas and Quirk, (2009) (Artwork by Liliana Cabrera.)

It is seen that the prelimbic cortex (PL) projects to the basal nucleus (BA) of the amygdala, which then excites the central nucleus (CE) of the amygdala, thereby promoting the expression of conditioned fear. Also, the BA also receives excitatory input from the lateral amygdala (LA), which also drives the expression of conditioned fear. In contrast, the infralimbic cortex (IL) excites a class of GABAergic inhibitory neurons known as the intercalated (ITC) cell masses. These neurons inhibit the CE, thereby inhibiting conditioned fear and promoting extinction. By comparison, PL and IL control cocaine seeking via their differential projections to the core and shell subdivisions of the nucleus accumbens (Figure 4.4).

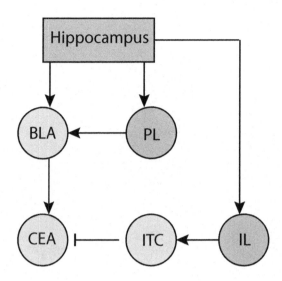

Figure 4.4 Neural circuit for context-dependent regulation of fear memory. PL rodent prelimbic cortex, IL infralimbic cortex, ITC, intercalated cells, BLA basolateral nucleus of the amygdala, CEA Central nucleus of the amygdala. (Based on Maren, et al (2013). [18[188] (Artwork by Liliana Cabrera.)

The dependence on context of fear memory involves a neural circuit that includes the hippocampus, the medial prefrontal cortex (specifically, the infralimbic cortex (IL) and the prelimbic cortex (PL)), and the amygdala (specifically, the basolateral amygdala (BLA), central amygdala (CEA), and intercalated (ITC) cells). The hippocampus projects directly to the BLA, and this projection may be crucial for the renewal of fear expression in response to an extinguished conditional stimulus. Indirect projections between the hippocampus and amygdala, via the medial prefrontal cortex, might also mediate the context-dependent expression of fear in response to an extinguished conditional stimulus. In particular, PL cortex projections to the BLA are involved in fear renewal. In contrast, IL cortex projections to ITC cells, which in turn inhibit CEA output, are involved in suppressing the expression of fear in response to an extinguished conditional stimulus [188].

Fear-related disorders

Some people are predisposed to the development of fear-related disorders from early life experiences, genetic backgrounds, and other risk factors. The expression of fear triggered by a memory of a traumatic event may serve to sensitize those who develop a mental disorder or psychopathology, resulting in an increased level of fear. Additionally, fear may become more generalized or associated with cues not related to the traumatic event in people who go on to develop a fear-related disorder [24]. In contrast, with increased resilience, fear responses to cues related to a traumatic event extinguish over time, and discrimination occurs between cues that are associated with the traumatic event and those that are not.

When a traumatic event occurs, people learn to fear the cues that are associated with the traumatic event, and the memory of the traumatic event consolidates over the subsequent hours and days. The expression of fear may come in several different forms, including flashbacks of the traumatic event, nightmares, avoidance of situations that trigger the memory of the traumatic event, and altered sympathetic responses such as being easily startled.

A model for the development of fear-related disorders

A model by Parsons, RG, Ressler, KJ (2013) of the development of fear-related disorders [24] describes how fear-related disorders develop:

Pre-existing sensitivity
(gene + environment)

↓

Learning of fear
(index traumatic event)

↓

Consolidation of fear
(hours to days following event)

↓

Expression of fear
(memories, nightmares, flashbacks,
Avoidance, sympathetic response, startle)

To reiterate, this development is different in people who are predisposed to the development of fear-related disorders from early life experiences, family genetics, and other risk factors. When a person experiences a traumatic event, the memory of the experience consolidates over the following days. Fear may be elicited from cues not associated with the traumatic event in those people who go on to develop a fear-related disorder. Also, resilience to fear responses to cues related to the traumatic event may decrease over time.

Treatment of fear-related disease

Fear-related diseases are prevalent in our society and are responsible for reducing many people's quality of life. Therefore, there have been many efforts devoted to finding effective treatments. In general, there are two different approaches that have been explored. The two approaches are behavioral techniques, such as psychotherapy, and the use of pharmacological treatments, such as medications.

Currently utilized treatments of fear and anxiety related disorders include antidepressants (targeting serotonin, norepinephrine, and monoaminergic dopamine pathways), GABA-acting benzodiazepines, beta-adrenergic receptor blockers, and cognitive-behavioral therapies. However, all of these treatments have limited efficacy. New direct and combined treatments specifically targeting known neural pathways underlying fear and anxiety are on the horizon.

The model shown below describes some current and proposed methods for future exploration (Figure 4.5).

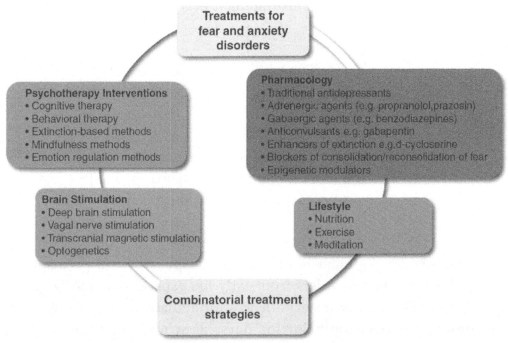

Figure 4.5 Current and promising new approaches to treatments for anxiety and fear-related disorders [22]. (Adapted from: Dias, Banerjee, Ressler, (2013) Towards new approaches to disorders of fear and anxiety, Current opinion in neurobiology, 23(3) pp346-352.)

Neurobiology of anxiety

Anxiety affects many body functions such as the autonomic nervous system, and anxiety can also increase the risk of diseases such as those related to aging [16]. Anxiety states are mediated by local and long-range connections between multiple brain areas. Some regions that have major roles in anxiety, such as the basolateral amygdala (BLA) and the anterodorsal bed nucleus of the stria terminalis (adBNST), mediate both anxiogenic and anxiolytic behavioral effects [185]. For example, activation of the BLA-to-ventral hippocampus (vHC) pathway causes an increased level of anxiety because it is anxiogenic. Whereas activation of the BLA-to-central amygdala (CEA) decreases anxiety because it is anxiolytic in nature.

Some other brain structures play important roles in connection with anxiety. These structures mainly consist of the basolateral amygdala, the ventral hippocampus (vHC), the central amygdala (CEA), the ventral BNST, and the ventral tegmental area (VTA). These structures can both enhance and suppress the behavioral expression of anxiety.

The two parallel pathways, the ventral BNST (vBNST)-to-ventral tegmental area (VTA) pathways, mediate either anxiogenic or anxiolytic behavioral outcomes. Also, large parts of the anxiety network remain to be characterized concerning cellular identity and functions as well as precise local and long-range connectivity using modern circuit-based approaches [185].

Sensory stimulation does not directly elicit the condition of anxiety as in fear. Fear-conditioning methods have indicated that anxiety is a form of a disease that is associated with deficits a person has with the extinction of a learned fear response. Anxiety states involve complex connections between many parts of the brain.

Emotional dysregulation

Studies have found evidence that emotional dysregulation is associated with generalized anxiety disorders or GAD [189]. These researchers found that the pre-frontal cortex (PFC) and the anterior cingulate cortex (ACC) inhibits the amygdala's activation in healthy individuals. These cortices act as a top-down regulation of the fear circuitry during an emotion regulation task.

The exaggerated sensitivity of different brain circuits to threats can help link anxiety with an increased risk for the diseases of aging. Such sustained threat perception is accompanied by the prolonged activation of threat-related neural circuitry and threat-responsive biological systems, including the hypothalamic-pituitary-adrenal axis, autonomic nervous system, and inflammatory responses. This prolonged activation ultimately can lead to elevated inflammation. Over time, the effects on central and peripheral systems may become chronic through structural changes in the central nervous system, altered sensitivity of receptors on immune cells, and accelerated cellular aging. Finally, such chronic elevations of inflammation can increase the risk for, while accelerating the progression of, the diseases related to aging [16].

Effect of anxiety on body systems

Fear may be acute or persistent. Anxiety is an unspecific sense of fear that lasts a long time. Anxiety is closely related to fear, which occurs as the result of threats that are perceived to be uncontrollable or unavoidable [21]. Anxiety can be described as a form of general and unspecific fear. Neurologists and psychiatrists regard some forms of anxiety to be a disease. Tremendous effort and money have been spent on treating such excessive fear.

Anxiety is a highly prevalent psychiatric disorder that often occurs together with depression, substance abuse, and many other psychiatric disorders. The disease is moderately inheritable, but genome studies have failed to identify genes that are noticeably associated with the disease [190]. Some anxiety disorders have been associated with various kinds of stress. Again, fear may be acute or persistent, and anxiety is an unspecific sense of fear that lasts for an extended duration.

Anxiety affects many fundamental systems in the body, and these changes may lead to severe medical conditions such as chronic inflammation, diabetes, and various diseases that are associated with aging. An integrative neurobiological model of the pathways mediating anxiety-related increased risk for diseases of aging is shown in figure 4.6.

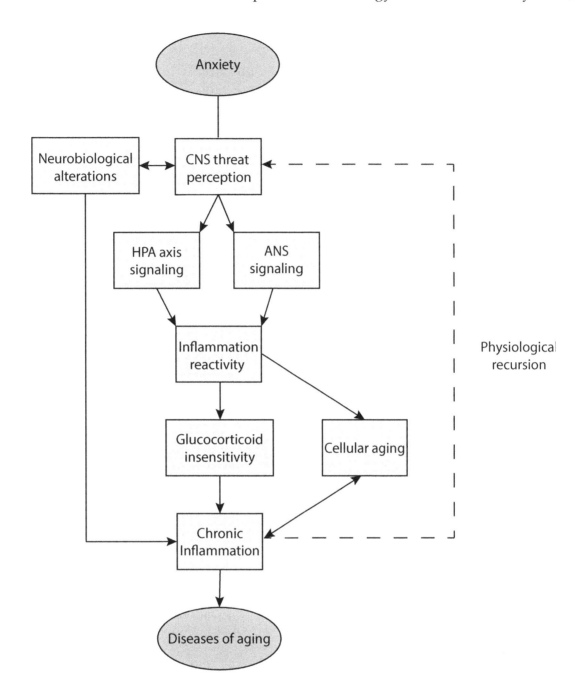

Figure 4.6 Integrative neurobiological model that is showing the pathways mediating anxiety-related increased risk for diseases of aging [16]. (Based on: O'Donovan et al. 2013.) (Artwork by Liliana Cabrera.)

The model in Figure 4.6 depicts how exaggerated neurobiological sensitivity to threat in anxious individuals leads to cognitive-behavioral threat responses characterized by a pattern of vigilance-avoidance, which ultimately results in sustained threat perception.

To review, such sustained threat perception is accompanied by prolonged activation of threat-related neural circuitry and threat-responsive biological systems, including the hypothalamic-pituitary-adrenal axis (HPA), the autonomic nervous system (ANS), and inflammatory response.

These systems and responses ultimately lead to an elevated level of inflammation. Over time, the effects on central and peripheral systems may become chronic through structural changes in the central nervous system (CNS), altered sensitivity of receptors on immune cells, and accelerated cellular aging. Therefore, such chronic elevations in inflammation can increase the risk for and accelerate the progression of, diseases of aging [16].

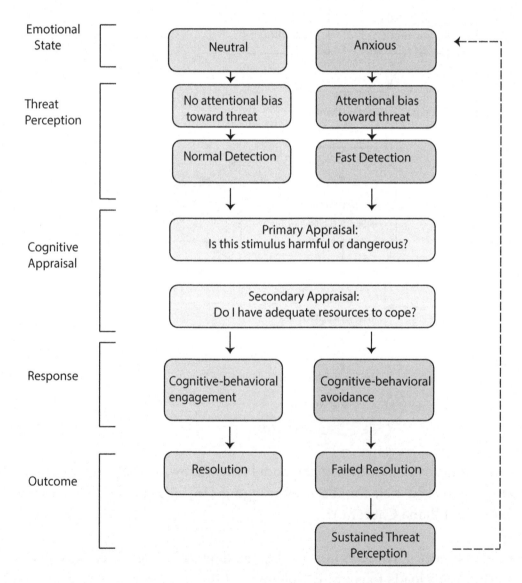

Figure 4.7 Overview of cognitive-behavioral responses to perceived threats in anxious and non-anxious persons. (Based on: O'Donovan, A., et al , 2013.) [16]. (Artwork by Liliana Cabrera.)

The response to perceived threats is different in anxious and non-anxious persons as shown in Figure 4.7. Anxious people show cognitive biases toward threatening information, which leads them to detect threatening stimuli, such as angry faces or predatory animals, more quickly than non-anxious individuals. Additionally, anxious people appraise both ambiguous and threatening stimuli as more threatening.

Anxious individuals also show a tendency to engage in cognitive-behavioral avoidance, which limits their ability to challenge inappropriate threat perception, confront and resolve threatening situations, and reshape expectations for the future. After a first and second appraisal, the anxious person has a cognitive-behavioral avoidance. In comparison, the non-anxious person has a behavioral engagement.

The ultimate result of this process is that the anxious person often fails to achieve a resolution of perceived threats, resulting in sustained threat perception.

Anxiety disorders

Anxiety disorders are one of the most common psychiatric illnesses. Close to 20% of the population suffers from an anxiety disorder of some kind in any given year [191] [5]. Anxiety disorders can cause such distress that it interferes with a person's ability to lead a healthy life. Anxiety disorders may be regarded as diseases.

Generalized anxiety disorder (GAD) features constant worry about a range of events, typically focused on the future. The psychological concepts related to anxiety and its treatment have been mapped onto behavioral processes such as adaptation and extinction. Anxiety has also been mapped onto their neural correlates [192]. There is a growing body of research in modern neuroscience regarding anxiety.

After a traumatic event, people learn the cues that are associated with the traumatic event, and this memory consolidates over the subsequent hours and days. The expression of fear comes in several different forms, including flashbacks of the traumatic event, nightmares, avoidance of situations that trigger a memory for the traumatic event, and altered sympathetic responses such as increased startle. The expression of fear, triggered by memories of the traumatic event, may serve to sensitize those who develop psychopathology, which results in an increased level of fear.

Additionally, fear may generalize to cues not associated with the traumatic event in those people who go on to develop a fear-related disorder. In contrast, with resilience, fear responses to cues related to the traumatic event extinguish over time, and discrimination occurs between cues that are associated with the traumatic event and those that are not [24].

Studies have found evidence that emotional dysregulation is associated with generalized anxiety disorders (GAD) [189]. Investigators found that the pre-frontal cortex (PFC) and anterior cingulate cortex (ACC) inhibits the amygdala's activation in healthy persons acting as a top-down regulation of the fear circuitry during an emotion regulation task.

However, there is a hypo-activation of the cortical areas in persons with generalized anxiety disorder (GAD), which leads to a deficit in the top-down control system during emotional regulation tasks [189]. Recent studies have shown that stronger amygdala-prefrontal connectivity predicts lower levels of anxiety, whereas stronger amygdala-prefrontal connectivity predicts effective regulation of emotions.

Hysteria

A pathological form of uncontrollable, unmanageable anxiety is often referred to as hysteria. The term hysteria has been used for more than two thousand years as a medical diagnosis, but it has been abandoned from modern use by the late nineteenth century and in the USA when the Psychiatric Association officially changed the diagnosis of hysterical neurosis, conversion type, to conversion disorder. The origin of the term hysteria may be from Hippocrates, who regarded hysteria to be a disease that affected the uterus. The Greek word hysteria means uterus. Therefore, the condition of hysteria was, at least at early times, referring to women. It was suggested that sexual intercourse would be an effective treatment because it would moisten the womb.

Mass hysteria is a form of contagious fear that may not have any real cause. Generally, modern medical professionals have abandoned using the term hysteria to denote a diagnostic category, replacing it with more precisely defined categories, such as somatization disorder. As stated before, in 1980, the American Psychiatric Association officially changed the diagnosis of hysterical neurosis, conversion type (the most extreme and useful type), to conversion disorder.

In most incidences with mass hysteria, the perceived fear is unlikely to materialize. Predictions of the end of the world were more prevalent in earlier years but still appear occasionally. Mass hysteria also occurs in response to rare diseases, to war, or to terrorism related events. Now it is driven by news media, and it is aimed at people who are defenseless because they do not understand the basis for the suggested diseases or the risk of other mass disasters.

Paranoia

Paranoia is a medical condition of fear, defined as a severe mental illness that causes a person to believe that other people are trying to harm the person without evidence or cause. To be more specific, it is a psychosis characterized by systematized delusions of persecution or grandeur usually without hallucinations.

Phobias

Phobias are unrealistic, irrational kinds of fear that are directed to specific situations or events. Phobias are a kind of anxiety disorder that are directed toward things or events that often pose little danger. Phobias are characterized both by predictive anxiety and acute flight responses. Also, phobias are often related to specific classes of stimuli, such as spiders or snakes.

Phobias are psychiatric medical disorders characterized by persistent fear of an object or situation. A person with a phobia will go to great lengths to avoid whatever the phobia may be about. There are many kinds of phobias, perhaps the most common phobia is acrophobia (the fear of heights), claustrophobia (the fear of closed-in places), xenophobia (the fear of strangers or foreigners), agoraphobic (the fear of open or public places), and social phobia (as discussed above) [28]. There are fewer known phobias such as the phobia of choking or Pseudodysphagia. Dental phobia, which may affect many people, is a phobia of pain. The phobia of pain is most pronounced in children, perhaps because there are indications that the non-classical sensory pathways are more active in children than in adults [64].

Dental phobia

Fear of pain is a common reason people avoid dental treatment, and many people even have dental phobia. A recent study found that people who did not have dental phobia had stronger and more widespread connections between the anterior cingulate cortex and the amygdala and the basal ganglia (specifically, the putamen, pallidum, and caudate nucleus) [193]. The people who had dental phobias had strong connections between the orbitofrontal cortex and the caudate nucleus. Previously, the basal ganglia were assumed to be involved in motor functions, but more recently it has become known that these nuclei have essential roles in many functions including emotions and cognitive functions.

If a person thinks about pain while viewing pictures of dental treatment, the activation of the orbitofrontal cortex (OFC) becomes enhanced. In a study of connectivity, Scharmüller and colleagues showed that this differential activation that occurs in people who are afraid of dental treatment is evidence of a different connectivity pattern compared with the control groups [194].

Specifically, these investigators found that people who are not afraid of dental treatment (the control group) had stronger and more widespread connectivity compared to the people with dental phobia. The investigators interpreted the pattern of connections to reflect successful emotion regulation in the control group. However, this was absent in the phobia group, and the phobia group showed coupling of the OFC and the caudate nucleus, which may be the neural correlate of associating pain with dental treatment.

Pseudodysphagia

The fear of choking, also referred to as pseudodysphagia or phagophobia, is a little-known phobia of eating that can have severe consequences by causing malnutrition or even starvation to death [195]. People with pseudodysphagia are afraid of eating and therefore avoid eating. What people are afraid of in other phobias can generally be avoided. However, some phobias like dental phobia and pseudodysphagia are more difficult to avoid completely. Going to the dentist can be an exception, because some dentists offer general anesthesia for dental work.

A neurobiological model of Social Anxiety Disorder (SAD)

Functional neuroimaging studies by Etkin and Wager have shown indications that the fear circuit is overactive in people with SAD [196]. More recently, a new model of the neurobiology of SAD based on studies of functional connectivity was presented by Brühl [197] (Figure 4.8 below). This study confirmed that the fear circuits (amygdala, insula, anterior cingulate and prefrontal cortex) are hyperactive in people with SAD. In addition to these structures, the medial parietal and occipital regions (posterior cingulate, precuneus, cuneus) showed signs of hyperactivity in SAD. Lastly, in people with SAD there was reduced connectivity between parietal and limbic and executive network regions.

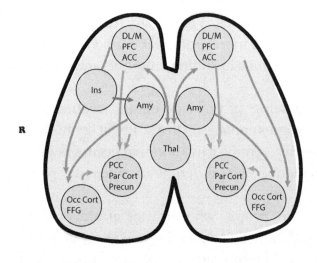

Figure 4.8 Neurobiological model of social anxiety disorder (SAD).

Abbreviations: DL/MPFC, Dorsolateral/Medial Prefrontal cortex; ACC, anterior cingulate cortex; Ins, insula; Amy, amygdala; Thal, thalamus; PCC, posterior cingulate cortex; Par, cort parietal cortex; Precun, precuneus; Occ, cort occipital cortex; FFG, fusiform gyrus; R, right; L, left. [197] (Based on: Brühl, et al, 2014.) (Artwork by Liliana Cabrera.)

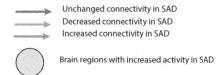

Unchanged connectivity in SAD
Decreased connectivity in SAD
Increased connectivity in SAD

Brain regions with increased activity in SAD

The illustration in Figure 4.8 shows the structures and connections that are more active in people with SAD than in control subjects (HCS). This study takes into account new developments in neuroscience [38] and especially points to the results of function connections (for "connectivity" see [82]). This study confirmed that hyperactivation occurs in the fear circuits (amygdala, insula, anterior cingulate and prefrontal cortex) in patients with SAD. This study [197] also mentioned that task-related functional studies revealed hyperactivation of medial parietal and occipital regions (posterior cingulate, precuneus, cuneus) in SAD. Additionally, reduced connectivity between parietal and limbic and executive network regions present an updated model of SAD adopting a network-based perspective. Therefore, the authors presented an updated model of SAD adopting a network-based perspective to studies of the neuroscience of SAD. This model was based on the result of a meta-analysis and review of a network-based perspective. The researchers propose that the disconnection of the medial parietal hub in SAD extends the current frameworks for future research in anxiety disorders.

Treatment of anxiety disorders

To reiterate, certain people are predisposed to the development of fear-related disorders due to early life experiences, their genetic background, and other risk factors. After a traumatic event occurs, people learn to fear the cues that are associated with the traumatic event, and this memory consolidates throughout the subsequent hours and days.

The preferred treatment of many disorders, including fear and anxiety, are to place patients on medication regiments of one kind or another. Many physicians generally prefer pharmaceutical products, even though other forms of treatment may have advantages. Other forms of treatment, such as physical exercise, have shown efficacy in treating mood related conditions such as fear and anxiety. Also, people with general stress have favorable effects from being physically active [23].

There is neuroscientific evidence of a positive effect, such as increased synthesis of brain-derived neurotrophic factor (BDNF) like when taking Omega 3 fatty acids (which also have many other beneficial effects), from physical exercise ([198, 199]). Additionally, the "recycling" of cell organelles is facilitated by physical exercise.

Threat-related neural circuitry includes the hypothalamic-pituitary-adrenal (HPA) axis, the autonomic nervous system, and inflammatory responses. Exaggerated neurobiological sensitivity to threats, thus, may be a treatment target for reducing disease risk in anxious individuals.

It was shown above that a person with anxiety reacts to fear in a different way than a person who does not have anxiety. Also, the neurobiology of sensitivity to perceived threats in a person with anxiety is different from that of a person who does not have anxiety.

Treatment of social anxiety

Figure 4.9 describes some methods to overcome social anxiety using different forms of cognitive behavioral therapy.

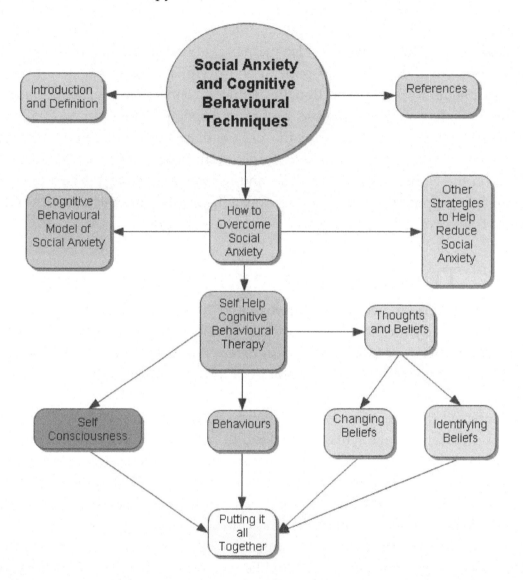

Figure 4.9 Cognitive behavioral techniques for the treatment of social anxiety (Adapted from Social Anxiety and Cognitive Behavioral Technique, Edward Griffin).

From Figure 4.9, it is seen that three routes may lead to overcoming social anxiety, (1) understanding the cognitive and behavioral model of social anxiety, (2) other strategies to help to reduce social anxiety, and (3) then using self-help cognitive behavioral therapy.

Again, this last suggestion regarding therapies of social anxiety has three different options. Finally, it is concluded that all these options can be put together to achieve a treatment that may be helpful and beneficial for people with social anxiety.

Chapter 5 Neurobiology of threats

Introduction

The presence of an environmental threat, in effect a stressor, leads to the activation of the brain's fear system, thus initiating the stress response in both the brain and the body. As reviewed in the preceding chapters, a fundamental structure of the fear system is the amygdala which is responsible for the detection of threat and the orchestration of stress responses in the brain and the body.

As mentioned in other parts of the book, the amygdala receives input from other systems of the brain, such as the sensory system and several parts of the cerebral cortex. Cells in the amygdala make connections with numerous systems of the brain, and the strength of these connections can change through activation of neuroplasticity.

There are several forms of threats and fear, and these different forms are likely to involve different structures of the brain which then respond in different ways. Behavioral and biological responses to threats depend on the activation of other brain areas than the amygdala, such as the bed nucleus of the stria terminalis (BNST), which coordinates autonomic and motor responses to threat. Additionally, the periaqueductal gray coordinates stereotyped defensive reactions to threat, such as immobility and panic ([16]).

Phasic fear responses to threatening stimuli are thought to be distinctly different from sustained, anticipatory anxiety toward an unpredicted, potential threat.

Generally, fear arousal initiated by an environmental threat leads to the activation of the stress response. The stress response is a state of alarm that promotes an array of autonomic and endocrine changes designed to aid self-preservation. The stress response includes the release of glucocorticoids from the adrenal cortex and catecholamines from the adrenal medulla and sympathetic nerves. These stress hormones, in turn, provide feedback to the brain and influence neural structures that control emotion and cognition. How this impacts fear conditioning can be described using the behavioral paradigm that has been widely used in studies of the neural mechanisms underlying the acquisition, expression, consolidation, reconsolidation, and extinction of emotional memories.

Several studies have shown that the nuclei of the amygdala and their connections to other parts of the brain, most importantly to the prefrontal cortex, play essential roles in the processing of information related to fear and anxiety. This is achieved mainly through these structures' role in memory. Also, these memory features influence the function of many other parts of the brain.

The amygdala plays a vital role in determining whether signals from the environment indicate whether or not the environment is safe or dangerous. The hippocampus and medial prefrontal cortex contribute to these functions as well. Signals from these structures can up-regulate or down-regulate the amygdala's sensitivity to threats.

Studies in humans have shown a clear relationship between the amygdala and fear [200]. Furthermore, studies of people with damage to the amygdala complex had difficulties in recognizing threats [94].

In this chapter, it is also discussed how the endocrine mediators of the stress response influence the morphological and electrophysiological properties of neurons in the brain areas that are crucial for fear-conditioning processes. These brain areas include the amygdala, hippocampus, and prefrontal cortex. The information in this review illuminates the behavioral and cellular events that underlie the feedforward and feedback networks which mediate states of fear and stress and their interaction in the brain [201].

Neural pathways of threat responses,

The neural pathways and neural processing of environmental or external information that activate a fear response are similar to those that are involved in eliciting a fear response. The input reaches the lateral nucleus of the amygdala from the sensory thalamus either directly from the dorso-medial thalamus (non-classical pathway) or from all parts of the sensory thalamus through the sensory cortices. Olfactory information reaches the central nucleus of the amygdala from the olfactory bulb. Information can also reach the lateral nucleus from the hippocampus, the entorhinal cortex, the prefrontal cortex, and the polymodal cortex (part of the association cortex). Pain and visceral information can reach the central nucleus from the sensory brainstem.

Information from the basolateral nucleus of the amygdala can reach cells in the prefrontal cortices, the ventral striatum, and the polymodal association cortices. Information from the central nucleus can reach the HPA axis and the hypothalamus. A freezing reaction is initiated from the periaqueductal gray (PAG) in response to signals from the central nucleus of the amygdala. The central nucleus also activates hormones that have modulatory functions such as norepinephrine, 5HT (serotonin), dopamine, and acetylcholine. The central nucleus of the amygdala can also activate the autonomic nervous system. Also, the central nucleus of the amygdala can change the balance between the activity in the sympathetic and the parasympathetic systems of the autonomic nervous system [201].

The amygdala is involved in a person's awareness of threats and their reactions to threats. Evidence shows that the amygdala is mainly involved in specific threats (Somerville [202]), whereas the bed nucleus is involved in uncertain or unspecific threats (Sink et al. [35, 203]). The nerve cells in the amygdala connect to many parts of the brain, and the connections to the prefrontal cortex are especially crucial for the expression of emotions such as fear.

The lateral nucleus of the amygdala is the main receiving nucleus for signals from sensory systems (see [1]), and the central nucleus is the main output nucleus that communicate reactions to threats such as freezing to other brain structures that cause increase in blood pressure and heart rate, and controls liberation of many different hormones.

So far, only a few published studies have focused on the separation of neural processes related to both phasic and sustained fear in a specific phobia. It has been suggested that first, conditions of phasic and sustained fear would involve different neural networks. Second, that overall neural activity would be enhanced in a sample of people with phobias compared to non-phobic participants.

To review, one of the earliest studies of the involvement of the amygdala in fear evoked by threats was the well-known study in monkeys by two German investigators, Klüver and Bucy, which was published 1939 [58]. In their studies of monkeys, they showed that damage to the amygdala nuclei was associated with several changes in the monkeys' behavior, including extreme tameness, signs of lack of fear, and the elevated threshold for threats (increased resilience). These investigators concluded that the reason the monkeys lost their fear was, at least partly, an interruption in the limbic pathways. This matter was discussed in detail in chapter 3.

Threat-related networks

Sensory signals that travel through the dorsomedial thalamus can reach the emotional brain directly from the thalamus via subcortical connections. Threats activate sensory systems that initiate actions quickly through the ascending, non-classical pathways which use the dorsal-medial thalamus. This pathway, the "fast and dirty" pathway, is fast and passes information directly to the amygdala for action. This pathway provides little processing and the signals in this pathway can place the body in an immediate alarm mode. Any sensory signal can elicit an alarm reaction through this pathway.

Sensory signals can also reach the emotional brain by traveling through the ventral thalamus and a long chain of cortical structures, also known as the classical pathways. Signals that travel through the classical pathways and undergo processing that can discriminate between different sensory signals. Thus, making it possible to distinguish between different kinds of threats.

The amygdala functions in concert with other brain regions, including the hippocampus and medial prefrontal cortex, which can up-regulate or down-regulate the amygdala's responses to threat. Moreover, the behavioral and biological responses to threats depend on activation of other brain areas, including the bed nucleus of the stria terminalis, which coordinates autonomic and motor responses to threat. These responses also depend on the periaqueductal gray, which coordinates stereotyped defensive reactions to threat, such as immobility and panic. Activity in this threat-related neural network is potentiated for individuals with anxiety disorders, as well as for persons exhibiting high levels of trait anxiety [16].

Each of the three principal subdivisions of the amygdala has extensive connections with the frontal lobe. These connections are essential for many of the features of fear, such as the fear of learning and the fear reflex.

Sensory pathways important for detecting suspicious matters and threats

Many parts of the brain are involved in the decision about suspicious matters, and there are two routes to actions on dangerous matters. The fastest route is designed for immediate defensive actions, it focuses on bodily responses and acts unconsciously. The other path is a slower route involving information processing and activation of consciousness making a person aware, feel the emotion, and comprehend its meaning. ("The Emotional Brain" by LeDoux.)

O'Donovan et al. (2013) presented a hypothetical model of the neural systems involved in detecting threats and regulating behavioral and biological responses to these threats [16]. The model shows the amygdala in the center with two-way connections to the hippocampus and the medial prefrontal cortex.

Involvement of the prefrontal cortex

Communication between the basolateral nucleus of the amygdala (BLA) and the medial prefrontal cortex (mPFC) is essential for the evaluation of threats and safety. Connections between the prefrontal cortex and subcortical structures activated by fear play an essential role in fear responses [204]. Recent animal studies have shown that lacking interactions between the prefrontal cortex and the basolateral nucleus of the amygdala can lead to higher levels of general fear and anxiety [205].

The prefrontal cortex is a part of the frontal lobe of the brain. The cells in this part of the brain receive input from the central nucleus of the amygdala. These cells in the prefrontal cortex also make connections to many other parts of the brain, and it can be regarded as a neural hub that influences the activity of many other brain regions. Two parts of the prefrontal cortex are especially important in connection with fear, the ventromedial and dorsolateral prefrontal cortices.

The ventromedial and dorsolateral prefrontal cortices

The dorsomedial prefrontal cortex is essential for coordinating brain activity that allows a person to take action. The dorsolateral prefrontal cortex connects to other brain regions such as the ventromedial prefrontal cortex, amygdala, and the insular lobe. The cells in these structures have a similar complex role in connection with predictable threats.

Dorsolateral prefrontal regions are associated with non-emotional sensory and motor areas. These areas, for example, are the basal ganglia and parietal cortex [206]. The dorsomedial prefrontal cortex serves as a neural hub that influences activity in other brain regions when threats are unpredictable.

Five visual fear signaling pathways to the amygdala

All five pathways have different lengths and consequently different latencies. These different lengths and latencies facilitate fear-relevant signals arriving at the amygdala at different times (Figure 5.1). Projections away from the amygdala have various behavioral effects by releasing norepinephrine via the locus coeruleus (LC) for alertness, upregulating attention, and boosting activity in recurrent loops. Amygdala connections to PFC can also initiate reappraisal, which can either reduce or increase the fear response depending on the perceived danger.

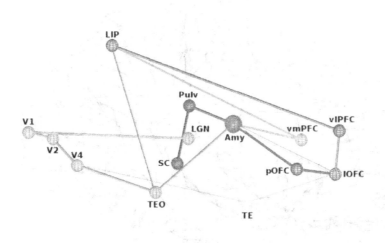

Figure 5.1 View of five visual fear signaling pathways to the amygdala. Pathways p1 is red, p2 is orange, p3 is orange to yellow at V4, p4 is orange to green at TEO, and p5 is orange to green at TEO, then blue at LIP [36]. (Reproduced from Silverstein, D. N and Ingvar, M (2015): A multi-pathway hypothesis for human visual fear signaling, Front Syst Neurosci. 9(9), p 101, following the Journal's rules.) [36].

The pathways for threat information and processing of threats are similar to those of fear and other emotions. Dual pathways to the emotional brain from sensory organs were initially observed in rats. Functional evidence was also observed in which applies to primates and specifically humans as well.

The regions of the limbic system are highly interconnected and function as a series of integrated parallel circuits that regulate emotional states. Each is densely innervated by the brain's monoaminergic systems. Noradrenaline (from the locus coeruleus (LC)), dopamine (from the ventral tegmental area, VTA), and serotonin (from the raphe nuclei (not shown)) are thought to modulate the activity of these areas. For details, see [36]. An overview of some of this connectivity is shown in Figure 5.2.

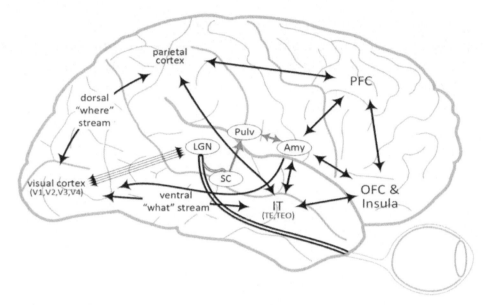

Figure 5.2. A more detailed overview of some brain regions involved in fear processing and some of the connectivity between them. This figure demonstrates some of the complexity of the brain circuits involved in fear processing [36]. Reproduced from Silverstein, D. N and Ingvar, M (2015): A multi-pathway hypothesis for human visual fear signaling, Front Syst Neurosci. 9(9), p 101, following the Journal's rules.) [36].

In humans, projections were found between the superior colliculus (SC) and the amygdala via the pulvinar *in vivo*. These projections were found using diffusion tensor imaging (DTI). Along this pathway, the SC is capable of image detection of fear-relevant stimuli at low spatial frequencies. A recent study found that neurons in the macaque pulvinar can respond selectively to the threat from snakes in 55 ms, which is likely too fast for a cortical route. It has also been found that the amygdala can be activated with low latencies from a fear-relevant stimulus in about 40– 120 ms, perhaps along the low route.

The role of limbic regions in resilience to the threatening nature of stressful situations

Brain imaging studies in humans, together with work in rodents and non-human primates, is now beginning to define the brain circuits that mediate distinct aspects of threats, fear, and other emotions under normal circumstances as well as other various pathological conditions [207, 208]. The field has identified several limbic regions in the forebrain. These limbic regions are highly interconnected and function as a series of integrated parallel circuits that regulate emotional states. An anatomical illustration of the similarity between the connections between sensory cortices, thalamus, and the amygdala is shown in Figure 5.3.

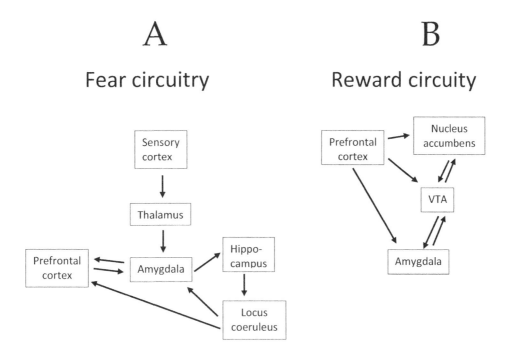

Figure 5.3 Anatomical differences between circuits activate in connection with fear and circuits activated in connections with reward (from [208]). (Based on: Feder, Nestler, and Charney (2009) Psychobiology and molecular genetics of resilience, Nature Reviews Neuroscience 10, pp446-457.) (Artwork by Liliana Cabrera.)

This figure shows some similarities between the neural networks in the limbic region of the brain that are involved in the fear and reward circuitries.

It is seen in Figure 5.3 A that fear-inducing sensory information is relayed through the thalamus (Thal) to the amygdala (Amy). The amygdala is particularly crucial for conditioned aspects of learning and memory, as is best studied in fear models.

The hippocampus has a crucial role in declarative memory, but it probably functions more broadly in regulating emotional behavior including fear [15]. The nucleus accumbens (NAc) is a crucial reward region that regulates an individual's response to natural rewards and mediates the addicting mechanisms of prevalently abused drugs (Figure 5.3 B). The prefrontal cortex (PFC), which is composed of multiple regions such as the dorsolateral PFC, the medial PFC, the orbitofrontal cortex, the anterior cingulate cortex, and others, has distinct but overlapping functions. The PFC is sometimes included in the limbic system and is essential to emotion regulation (Figure 5.3 B). PFC regions provide top-down control of emotional responses by acting on both the amygdala and the NAc (Figure 5.3 A and B). The limbic regions are also part of integrated circuits that mediate responses and behavior [208].

Other regions are important for fear and reward learning that are not shown in the respective circuits in Figure 5.3. For example, the NAc also regulates responses to fearful stimuli, and the hippocampus regulates responses to rewarding stimuli as well. The functional status of all of these circuits has significant implications for resilience to stressful life events.

Discrimination between dangerous objects and harmless objects

What is interpreted as fearful depends on the person because different people are afraid of different things. For example, if a person who is afraid of snakes see a snake the visual information will be interpreted in a way that will create a sense of fear in the structure of the brain where awareness of the image of a snake was created. In another part of the brain, that same information will create a fear reaction. If a person who is not afraid of a snake sees the same snake, it will cause the same awareness of a snake. However, the structure in the brain that is devoted to fear perceptions will not have the same activation.

Seeing or hearing something that might be threatening will activate many structures in the brain. The first action is to determine if the signal that is received through the sensory systems indicate an imminent danger or not. There are two main pathways for sensory signals, the classical and the non-classical pathways. The non-classical pathway performs little processing but is the fastest pathway. Also known as the "fast and dirty" pathway. The classical pathway performs extensive procession of sensory signals but is slow. Also known as the "slow and accurate" pathway.

A detailed description of the biology of detection of threats that involves the classical ascending sensory pathways and the non-classical pathways, direct connections from the thalamus to the lateral nucleus of the amygdala, are shown in Figure 5.4. The pathways that use the ventral thalamus, the classical ascending sensory pathways, take a longer time than the non-classical pathways.

However, the classical pathway makes it possible to distinguish between what is a threat and what is not, such as the discrimination between a stick and a snake (Figure 5.4).

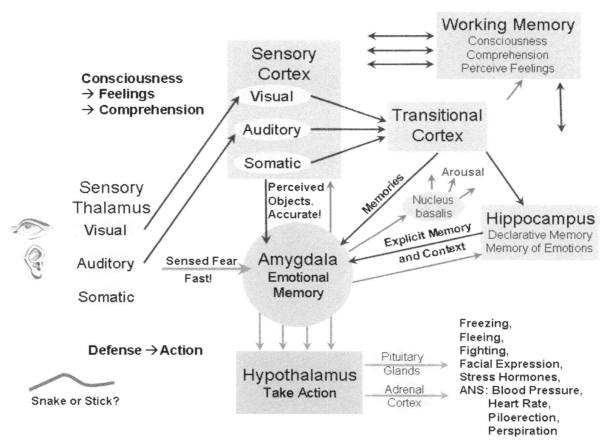

Figure 5.4 Neurobiology of a threat. The distinction between a snake and a stick is illustrated.

The description of the neural circuits involved in deciding whether a sensory (visual) message or stimulus indicates that an object is harmless or dangerous. It is seen how the information from the thalamus can take two ways to the center of the emotional brain; information can take a short route directly from the dorso-medial thalamus to the lateral nucleus of the amygdala (the non-classical pathways), or the information can take a long route from the ventral thalamus through a chain of cortical cells to the lateral nucleus of the amygdala (the classical pathways). In the classical pathways, the information is analyzed on its way from sensory receptors to the amygdala.

This makes it possible to distinguish between objects that are dangerous or harmless, such as a stick. Note the extensive connections to memory circuits in the brain. Threat perception leads to the activation of the hypothalamic-pituitary-adrenal (HPA) axis which then leads to an increased release of the glucocorticoid hormone, cortisol, from the adrenal glands (Figure 5.5).

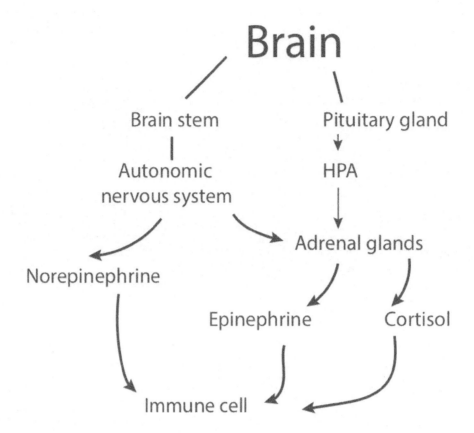

Figure 5.5 Illustration of the pathways linking threat-related neural activity in the amygdala, medial prefrontal cortex, and hippocampus with elevated inflammation (HPA-cortisol path) [16]. (Based on: O'Donovan, Slavich, Epel and Neylan 2013.) (Artwork by Liliana Cabrera.)

The adrenal glands are activated as a part of preparing the body for the fight or flight response in an emergency. This means that fear can cause this gland to secrete adrenaline into the bloodstream, and that secretion will elevate the blood pressure, increase heart rate, and cause appropriate changes in the internal milieu of the body. This combination may induce inflammation.

Threat perception of danger activates the sympathetic arm and deactivates the parasympathetic arm of the autonomic nervous system (ANS). This activation and deactivation lead to an increased release of the catecholamines, epinephrine and norepinephrine (solid lines). The pattern of activation and deactivation is accompanied by increased synthesis and release of pro-inflammatory cytokines. These cytokines include interleukin-1β (IL-1β), interleukin-6 (IL-6), and tumor necrosis factor-α (TNF-α).

Binding of these factors to receptors on immune cells regulates gene expression, including the expression of genes for pro-inflammatory cytokines. Thus, the effects of the HPA axis and the ANS on the immune system depend on the expression of immune cell receptors for cortisol and catecholamines, as well as the release of these hormones.

The glucocorticoid receptor (GR) appears to be down-regulated in response to the threat, thus limiting the anti-inflammatory effects of cortisol. Although there are complex bidirectional relationships between the various factors in this model, threat perception ultimately leads to elevated inflammation [16].

Crucial structures involved in creating the reactions to fear

In addition to the amygdala, the thalamus, the hypothalamus, and the hippocampus, the frontal and the temporal lobes play essential roles in creating the reactions to fear.

The thalamus acts as a giant switchboard that directs information to other parts of the brain. The fight-or-flight reaction is generated in the hypothalamus. From there, signals are sent to the adrenal glands, which then release stress hormones. The hippocampus, the sensory cortices, and the amygdala are the areas of the brain that established situational and emotional context. Also, these areas officially deem a situation as fearful. At the hypothalamus, fear-signaling impulses from the central nucleus of the amygdala activate both the sympathetic nervous system and the modulating systems of the HPA axis.

Frontal and temporal lobes are higher cortical areas where experiences of dread occur. These areas are also where the release of chemicals like dopamine that can cause panicked, irrational behavior occur.

The way fear may place the body in an alarm mode

Activation of the autonomic nervous system causes stress that wears on the body. There are now research results that indicate that stress can increase the risk of serious diseases related to the aging process. This can happen through the promotion of the inflammatory processes which may increase the risk of serious diseases such as some forms of cancer [16, 209].

Stress and diseases

Fear and anxiety are common causes of stress. Activation of the stress response is one of the risk factors for many diseases including the metabolic syndrome caused by insulin resistance that is related to obesity and diabetes type 2. Together with overeating and of the wrong things, and low physical exercise it is a risk factor for many other diseases such as cardiovascular diseases and ischemic strokes, various forms of cancer, dementia including Alzheimer's disease [209].

Chronic stress can lead to the development of metabolic syndrome and many other diseases. Activating the stress system affect the function of several organs including the immune system as illustrated in Figure 5.6.

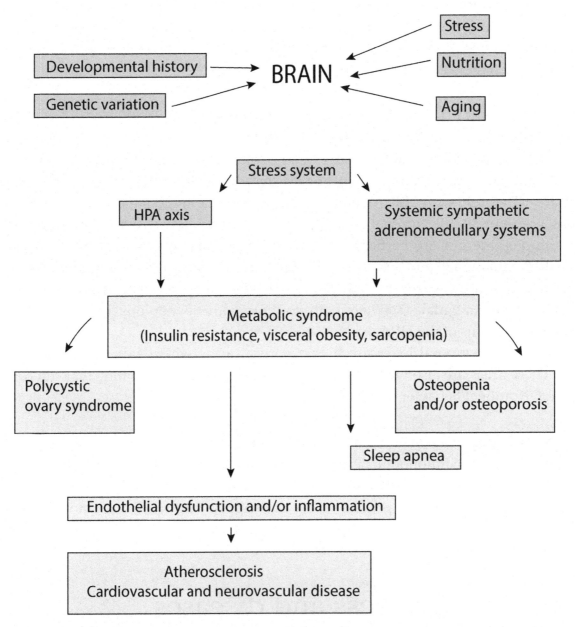

Figure 5.6 Steps in the developments of pathologies from chronic stress. (Based on Chrousos, 2009.) [209] (Artwork by Liliana Cabrera.)

Acute stress may trigger neuropsychiatric manifestations such as anxiety, depression, executive and cognitive dysfunction, cognitive dysfunctions, panic attacks, and psychotic episodes as well as body signs such as allergic reactions, eczema, urticaria, migraines, hypertensive or hypotensive attacks, different types of pain, gastrointestinal syndromes.

Lasting effects of trauma on the brain, showing long-term dysregulation of norepinephrine and cortisol systems and vulnerable areas of the hippocampus, amygdala, and medial prefrontal cortex [210] affect many organ systems of the body. The hypothalamic–pituitary–adrenal, HPA axis is activated by stress signals that travel from the paraventricular nucleus (PVN) of the hypothalamus to the pituitary gland and then to the adrenal glands for glucocorticoid release.

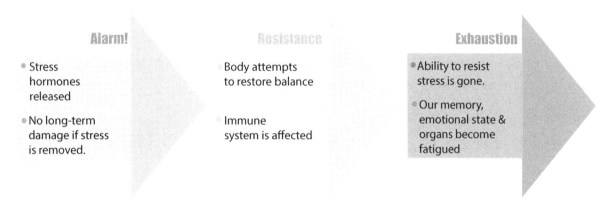

Figure 5.7 Sequence of effects of repeated activation of stress. (Artwork by Liliana Cabrera.)

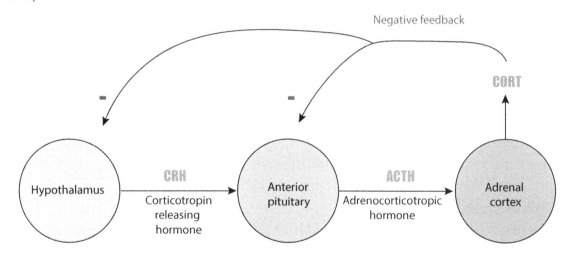

Figure 5.8 The HPA axis and its feed-back loops. (Artwork by Liliana Cabrera.)

The role of the visceral system in fear

Mood-related events, especially those of fear, activates many parts of the visceral system. These mechanisms were discussed in chapter 3, and Figure 3.31 showed the routes sensory signals take to reach the visceral system of the brain. The visceral system of the body can elicit activity in the autonomic nervous system which serves as a part of the visceral system's effector organs.

Information derived from the environment or recalled from memory is available to the amygdala and orbitofrontal cortex. Specifically, in the orbitofrontal cortex, information is sent to the hypothalamus, the basal forebrain, and nuclei in the brainstem tegmentum. A response from the viscera reaches the anterior, and the anterior cingulate cortex is the site of a second-order mapping [117].

Resilience against reacting to threating situations

Having a resilience to reacting to threatening situations has been associated with a tendency to perceive potentially stressful events in less-threatening terms and to remain optimistic about the ability to cope with stressors [207]. Social competence and the ability to harness social support have also been linked to better mental well-being and better physical health. Also, increased social support has buffering effects on mental and physical illness, and social support fosters adaptive coping strategies [207].

The neural processing of the signs of threats, which occur in different parts of the brain when a person experiences fear, is affected by many factors. Some of these are hormones, such as estrogen, and others are stress related. Also, many factors can modulate the function of the amygdala and alter the functional connections between the amygdala and the prefrontal cortex. These alterations can affect the reaction of resilience to the threatening situations [207].

Resilience has been linked to being able to perceive stressful events in less threatening ways, thus promoting adaptive coping strategies. These strategies along with one's cognitive reappraisal ability allows individuals to re-evaluate or reframe adverse experiences in a more positive light.

Many hardy individuals cite acceptance as a critical ingredient in their ability to tolerate highly stressful circumstances. Additionally, acceptance has been described as a common trait among highly successful learning-disabled adults and survivors of extreme environmental hardships and life-threatening circumstances [207].

The administration of various drugs that affect the function of the amygdala, nuclei can affect the reaction to a threatening situation at the same time as it affects a person's fear and anxiety.   The benzodiazepines, beginning with diazepam (Valium), were primarily aimed at enhancing inhibition in the amygdala by enhancing gamma-aminobutyric acid (GABA) receptors.   Benzodiazepines are GABA$_A$ receptor agonists. GABA is an essential inhibitor in many parts of the brain, and the amygdala nuclei are exceptionally rich in this receptor.   Therefore, the amygdala is especially affected by the administration of benzodiazepine such as diazepam and alprazolam (Xanax).

The administration of other drugs can also affect the function of the amygdala.   For example, the administration of the active component of cannabis, tetrahydrocannabinol or THC, is a current topic of research.   This is the component of cannabis that causes euphoria and a "high", and it can affect the function of the amygdala and thereby modulate anxiety [211].   It may achieve this by altering the functional connections between the amygdala and the prefrontal cortex. Gorka and his co-workers [211] studied these connections in humans using fMRI and found indications that THC enhanced connections between the basolateral nucleus of the amygdala and the rostral anterior cingulate/medial prefrontal cortex [211].

THC has the potential to reduce threat perception or enhance socio-emotional regulation.   A study in humans found that THC significantly reduced amygdala reactivity to threatening social signals but did not affect activity in the primary visual and motor cortex.   This is in good agreement with the notion that THC and other cannabinoids may have an anxiolytic role in central mechanisms of fear behaviors. This opens the possibility for a novel therapeutic strategy that target the cannabinoid system because it may have a beneficial effect for various kinds of emotional disorders [212] [211].

The other component of cannabis, CBD or cannabidiol, does not produce euphoria or a "high", but CBD has many different medically beneficial effects such as suppressing epileptic seizures, controlling of pain, and more [213].   Experiments with CBD demonstrated a non-activation of the neuroreceptors CB1 and CB2, but displayed a strong interaction between CBD and the 5-HT1 neuro-receptor was found [214].

There are receptors in the brain that specifically sense CBD and THC. These receptors are CB1, CB2, TRPV1, GP55, and others. There are also systems in the brain that can synthesize CDB and are known to control emotional behavior, mood, sleep, stress, irritability, fear, and even the sensation of "craving". These systems are located in the prefrontal cortex, amygdala, hippocampus, nucleus accumbens, and periaqueductal gray (PAG) of the midbrain.

In an early study on the effect of cannabis in humans, it was shown that CBD blocked the anxiety promoted by THC, thus indicating antagonism between the two [215]. Other studies have shown indications that CBD can reduce the symptoms of anxiety. There are also indications that CBD can reduce cognitive impairments in people with anxiety [216]. Other studies have found a beneficial antidepressant effect of CBD [214].

As mentioned above, it is now evident that symptoms of a disease may not only be caused by some parts of the brain that are malfunctioning but altered functional connections may also cause symptoms. Indirect projections between the hippocampus and amygdala via the medial prefrontal cortex might mediate the context-dependent expression of fear in response to an extinguished conditional stimulus.

Circuits involved with the extinction of fear

Structures involved explicitly with fear extinction

Fear expression and fear extinction are opposite processes, and this is reflected in the connections in the brain in these two situations (Figure 5.9). Specific connections in the brain are different during fear expression and fear extinction, especially the connection between the prelimbic (PL) structures and the basolateral nucleus of the amygdala (BA) as well as the connections between the infralimbic structures (IL) and the intercalated (ITC). The intercalated (ITC) cells of the amygdala are a group of GABAergic neurons situated between the basolateral and central nuclei of the amygdala which are important for inhibitory control over the amygdala [217].

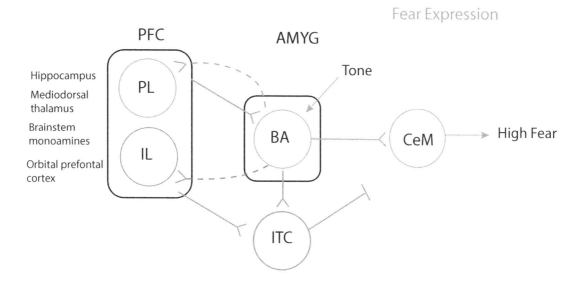

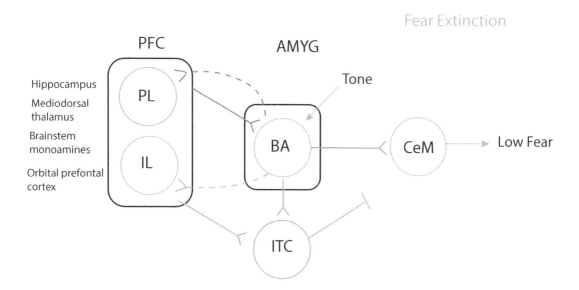

Figure 5.9 Prefrontal control of fear expression and extinction. During fear expression (left) and extinction (right) of auditory fear conditioning, tone responses from the amygdala (amyg) get integrated by the prelimbic (PL) or infralimbic (IL) prefrontal cortex with converging information from diverse sources, such as the hippocampus (Hipp), brainstem monoamines (Bstm), mediodorsal thalamus (MD), and orbital prefrontal cortex (OFC) to determine whether or not to produce a fear response [217]. CEm, medial central amygdala; LA, lateral amygdala   (Based on Sotres-Bayon, and Quirk, 2010.) [217]. (Artwork by Liliana Cabrera).

Fear excitation involves PL projections back to the basal amygdala (BA), whereas fear inhibition involves IL projections to amygdala-intercalated cells (ITC). In turn, the BA excites neurons in the medial division of the central nucleus of the amygdala (CeM) to produce fear responses, while ITCs inhibit these amygdala output neurons thereby inhibiting fear responses. Thus, the same conditioned stimulus, such as a tone, signals either high fear (red) or low fear (green) states in the appropriate circumstances. The extinction of fear is mediated by different circuit elements within the same structures. Input from the infralimbic cortex (IL) to the BA and the intercalated (ITC) cells is instrumental in dampening fear output from the lateral central amygdala (CEl) nuclei to the hypothalamus (HYP) and the periaqueductal gray (PAG). The identity, connectivity, and function of the critical forebrain-to-brainstem fear pathways are not fully understood and in need of being characterized by modern circuit-based approaches.

During fear expression and extinction, left side and right side in Figure 5.9 respectively, of auditory fear conditioning, tone responses from the amygdala (amyg) get integrated by the prelimbic (PL) or infralimbic (IL) prefrontal cortex with converging information from diverse sources such as the hippocampus (Hipp), brainstem monoamines (Bstm), mediodorsal thalamus (MD), and orbital prefrontal cortex (OFC). This integration of information determines whether or not to produce a fear response.

Fear inhibition during extinction depends on the bidirectional connectivity between the ventral part of the prefrontal cortex and the BLA (blue lines, PL in Figure 5.9). This connectivity ultimately regulates central amygdala activity to reduce conditioned fear responses [177]. During fear expression (red lines), fear-related contextual information is relayed from the hippocampus to the amygdala. The bidirectional loop between the BLA and the dorsal part of the prefrontal cortex allows updating of fear-related information, which is further transmitted to the central amygdala. From the central amygdala, the information goes to the periaqueductal gray for fear expression.

The context-dependence of fear memory involves a neural circuit that includes the hippocampus, the medial prefrontal cortex (specifically, the infralimbic cortex (IL) and the prelimbic cortex (PL)), and the amygdala (specifically, the basolateral amygdala (BLA), the central amygdala (CEA), and intercalated (ITC) cells).

Studies of the extinction of fear are essential not only as a possible way to treat fear-related disorders, but also because fear is a part of post-traumatic stress disorder (PTSD) (see Dejean [50, 177, 186]. Imaging studies in rodents have shown evidence that the prelimbic (PL) cortex and the dACC are associated with high fear expression, whereas the IL and vmPFC are associated with fear suppression.

Structures that are specifically involved in fear extinction, which is the ability to learn to suppress the experience of fear, are basal amygdala nucleus (BA), the ventral hippocampus (vHC), the prelimbic cortex (PL), the infralimbic cortex (IL), the intercalated (ITC), the lateral central amygdala (CEl), the hypothalamus (HYP), and the periaqueductal gray (PAG).

There are reciprocal connections between the basal amygdala nucleus (BA) and the ventral hippocampus (vHC), as well as between the BA and the PL. In turn, central nuclei of the amygdala project to hypothalamic and brainstem centers to promote fear behavior.

The top-down control of the amygdala by the hippocampus and PFC might result in the increased activation of the amygdala, as is observed in persons with PTSD. The end result of these neuroanatomical alterations is increased stress sensitivity, generalized fear responses, and impaired extinction. Other regions, including the anterior cingulate cortex, the orbitofrontal cortex, the parahippocampal gyrus, the thalamus, and the sensorimotor cortex, also play a secondary role in the regulation of fear and PTSD.

Much is now known about the mechanisms underlying the acquisition and subsequent extinction of conditioned fear. Also, it is generally accepted that fear extinction may be a useful model in understanding the underlying mechanisms in the pathophysiology of anxiety disorders. Additionally, the fear extinction model may also be a good model for current therapies that treat these disorders. It is, however, poorly known how men and women may differ in their biology of fear and its extinction. Estradiol (E2), a primary gonadal hormone, may modulate the mechanisms of fear extinction, and Estradiol may mediate some of the sex differences observed in psychiatric disorders [167].

The reciprocal modulation of gonadal hormones and stress response neurobiology may underlie the sex differences and the influence of gonadal hormones in fear extinction and anxiety disorders. These systems are also indirectly connected to the hippocampus, mPFC, and amygdala, which are critical regions in fear circuitry that are affected by stress and estrogen [146]. Maeng and Milad have discussed the interconnection between stress and fear mechanisms [146]. They found in reviewing the literature that women, in general, are more vulnerable to stress and fear-based disorders, such as anxiety and post-traumatic stress disorder. However, they found that the neural basis of these sex differences is unclear. These authors concluded that the studies of the cross-talk between neural systems that control fear and the neural systems that control stress are beginning to shed some light on the neuroscience of the interaction between fear, stress, and how it is affected by female reproductive hormones.

Chapter 6 Minimizing the risk of harm

Introduction

There are many forms of harmful situations and diseases in a modern society that can cause damage to the human body or even result in death. Recent research has increased the arsenal of remedies and methods for reducing the risk of harm caused by trauma and diseases. Some of these remedies are in the hands of individuals, while other solutions are in the hands of the planners and engineers of different facilities and equipment. Similarly, some of these advancements are in the hands of government offices that create regulations which are aimed at reducing the risk of various harms and diseases. The scientific advances that have made that possible have also reduced the fear and anxiety that many people experience from the associated risk of getting a disease that has no known prevention or treatment.

Many remedies are now available that can reduce the risk of the occurrence of certain diseases or other harmful events. An example of a preventive means that have saved many people's lives and more people from trauma is the seat belts in personal automobiles. Also, built-in safety devices in automobiles have made it less of a risk to travel in a private car. Through technological advancements, air travel has become extremely safe. Additionally, laws regarding the vaccination of children, rules about putting additives in food, and regulations on how food is handled have saved and protected numerous people from diseases. Now, the risk of acquiring some disease can be reduced by known means of prevention. However, economic factors often work in opposition because the means that can reduce the risks of harm usually cost money.

It is also important to consider if a person adheres to the available knowledge on a certain risk and how a person chooses between different options that have a different degree of risk. Logically, choosing a less risky option is a great alternative to minimize the risk of harm. A person's choice between options of different risk levels should be based on the value of the estimated risk. For example, various options for travel may have significant differences in risk, such as the choice between flying commercial or driving a personal automobile.

Comparing the risk associated with different kinds of harm is essential for minimizing the risk of harm. However, people place different values on different forms of risks, and the risk value one places on a particular situation or disease is not always guided by the knowledge about the likelihood that a person will suffer harm from that particular activity.

Minimizing risk often depends on making the optimal choice between available options. In medicine, there are plenty of options and remedies from which treatment can be selected. In many other common situations, people make choices based on various criteria and reducing the risk of harm is one such criterion. Choosing the optimal achievement is another. The decisions often made are based on uncertainty. Frequently, the task is to minimize harm and optimize convenience. These decisions are made under uncertainty.

Management of risks, therefore, involves making decisions under uncertainty. To be able to make those decisions effectively and rationally, it is necessary to have measures risks. In business and engineering, it is common to assign numbers to risks. A French mathematician, Blaise Pascal (1623-1662), laid the basis of probability theory, and Jacob Bernoulli introduced the measurement of risk around 1705 [14]. Assigning quantitative values to various risks has made it possible to manage risk quantitatively, instead of using ones "gut feeling" in making decisions.

The situation is different in the field of health management and for the treatment of diseases. The use of quantitative risk management techniques is nearly nonexistent for matters that concern risks of deaths, bodily harm, and other impairments of health. Yet a person's well-being should be more important than business matters. While it is possible to buy insurance that covers the cost of a particular medical treatment, no insurance protects a person against becoming ill or the harm that may occur from an illness in connection to the medical and surgical treatments. This is why it is so important to take preventive action to reduce the risk of injuries or diseases. These matters are discussed below in several parts of this chapter.

Placing different values on different kinds of harm

How often an adverse event occurs and how many people are involved is a measurement that is used as the basis for the estimation of the likelihood that a particular person will be affected by the adverse event in question. However, people are often assigned a value to most forms of harm based on different criteria. The way an adverse event is caused plays a major role in how harmful a person regards a particular event to be.

The value that a person assigns to different kinds of harm is often minimally related to the real severity of the risk of specific harm, but the value is more affected by how many people are involved in a single incident. Also, a person's reaction to the risk of harm to a person's body is minimally related to the likelihood that the adverse event in question will ever affect the person.

For example, the news media reacts in different ways to adverse events, and that gives people a yardstick with which to assess the extent and consequences of those adverse events in general. These measurements are rarely related to the implications the adverse event will have on a person.

While there is a very small risk that an individual would die from an airline accident, a person has a significant risk of being involved in an automobile accident. Similarly, there is a very small risk that a person would get hurt by a terrorist attack provided that terrorism stays at its present level. For an individual, it does not matter whether he/she dies from a mundane error in a hospital or from a spectacular event that will appear on the front page of national papers. The outcome of death is the same.

School shootings are evaluated strongly compared with other harmful events. The number of people affected in school shootings can cause significant changes in the way matters are handled, whereas harm that affects many people can occur without any action taken. This happens even in cases where the remedy for practical actions are present. An example is a dramatic increase in death from influenza from an average of 20,000 deaths to 80,000 deaths in 2018 that had little coverage by news media. A rare disease that killed two people on the same day will be on the front page of major newspapers, but an annual increase of 60,000 deaths from the influenza virus was barely mentioned by the news media.

Accidents related to commercial airlines get much more coverage by the news media than other accidents despite being exceptionally rare and only affecting relatively few people. While there is a minimal chance that a person in the USA will be affected by terrorism, the topic is still covered extensively by the news media which causes many people to fear the risk of being harmed by terrorism. Automobile accidents, also referred to as motor vehicle accidents or MVA, are only mentioned by the news media when there are many people involved or when the circumstances are exceptional. MVA are by far more likely to affect a person than accidents of commercial airlines.

The number of deaths from preventable errors and mistakes in the health care system in the USA is not known accurately. Two large studies [218, 219] found 44,000-98,000 deaths due to preventable causes from various mistakes and forms of errors during one year in US hospitals. The smallest estimation (44,000) is approximately the same as all deaths in automobile accidents (43,000). More recent studies [220] show even larger numbers of death from medical and surgical treatment.

It was reported that approximately 12,000 patients died per year from complication from operations that provided no benefit for the patient, 7,000 deaths occurred per year from medication errors in hospitals, 20,000 deaths from other errors in hospitals, 80,000 deaths from nosocomial infections in hospitals, and 106,000 deaths per year occurred from adverse effects of medications.

These numbers of patients who lost their lives are only slightly less than that of Alzheimer's disease. The fact that two well-performed studies produced results that are so different, 98,000 versus 44,000 deaths, show how difficult it is to obtain reliable information about the extent of risks in medical care systems [221]. Not even the number of deaths can be verified with reasonable accuracy, but it is even more difficult to find out the number of patients who have been injured and survived their injuries. Nevertheless, such incidences are essential to consider. For example, patients can experience injuries in a surgical operation that might cause lifelong suffering, such as the restriction of movement (paresis or paralysis), severe pain, or the loss or impairment of one or more senses. Unfortunately, the number of such incidents is not readily known. Also, the risk is not the same for all patients, and it depends on what disease a patient has as well as the patient's risk factors. Older adults would generally have a higher risk than younger ones, or patients who have a more risky treatment or operation [222].

Other studies have shown an estimated 3-4 percent of hospitalized patients experience adverse effects of preventable injuries during hospitalization [219, 223]. A journal article from the Annals of Internal Medicine points out that the problems in medicine are more widespread than just safety and systemic problems in the healthcare system, and these problems may cost hundreds of thousands more lives every year than do lapses in safety [224]. This article points to other kinds of lapses in providing adequate care. In fact, the problems that are pointed out cost more lives and cause more injuries than simple mistakes and errors. For example, the National Committee for Quality Assurance (2003) has estimated that 57,000 deaths occur annually because of shortcomings in delivering recommended care.

Furthermore, it has been estimated that 700,000 individuals die every year because of easily preventable causes, such as not been vaccinated against common disorders and a failure to reduce their risk factor of obesity through a healthy diet [225]. It is not directly caused by a failure in the medical system, but it also deserves to be mentioned that an estimated 400,000 individuals die every year from tobacco use. About half of all Americans who continue to smoke cigarettes will die at the hands of the habit. Nearly one of every five deaths is related to smoking cigarettes. Smoking cigarettes has killed more Americans than alcohol, automobile accidents, suicide, AIDS, homicide, and illegal drugs combined (Surgeon General's Report on "The Health Consequences of Smoking" 2004).

Increasing the use of preventive care could reduce the frequency of being hospitalized. Thus, one could reduce the risk of harm in health care, and hospitals would only be treating patients who would benefit from the treatment. Also, people would reduce their risk of complications from surgical operations, examinations, and tests that do not provide benefit to the patient. Ultimately, following these principles would also reduce a patient's fear and anxiety.

While the extent of the adverse effects in hospitals is poorly known, it is even less well known what occurs to outpatients. Outpatient adverse drug effects are difficult to estimate, but some authors have arrived at a number close to 200,000 deaths annually [220] [222]). One study shows that between 4 percent and 18 percent of outpatients experience adverse effects from their treatment. These adverse effects caused 116 million extra physician visits, 77 million extra prescriptions, 17 million emergency room visits, 8 million hospitalizations, 3 million long-term admissions, 199,000 additional deaths, and $77 billion (USD) in extra costs [220, 222].

The fact that outpatients must control and administer the medications that are prescribed by physicians and surgeons can lead to overdosing. For example, if a prescribed pain medication does not relieve a patient's pain with the prescribed dosage, the patient may increase the dosage to get the anticipated effect without consulting the physician who prescribed the medication. If the medication is an opioid, taking two or three pills, instead of the prescribed one pill, may lead to death because opioids in high dosages cause respiratory arrest. There is also the concern of a patient becoming addicted if they take more than prescribed. If the medication is acetaminophen, the common brand name being Tylenol, increasing the prescribed dosage may exceed the limit where there is a high risk of liver failure (4,000mg in 24 hours). People consuming too much acetaminophen causes more people to get liver failure than any other cause [226-229].

In order to use this research and these developments to minimize risks, it is necessary to understand risk assessment and how to make decisions under uncertainty. Preventative actions to reduce the risk that something unwanted occurs can reduce a person's fear and anxiety.

Many forms of treatment are unpleasant and painful. Also, most forms of treatment have negative side effects. However, preventative medicines reduce the need to treat many diseases in the first place, such as infectious diseases, because of the availability of vaccinations. Now, there are also many other examples of ways to reduce the risk of injury and diseases from many different sources. In addition to reducing the risk of diseases that may or may not be curable, it also prevents the suffering from the side effects of treatment.

The power of preventive means for treating diseases

So far, preventive means have shown an ability to reduce the risk of acquiring many diseases. Vaccinations have reduced the risk of many infectious diseases to the extent of possible elimination of a disease. Preventions, in the form of change in lifestyle, have reduced the risk of acquiring many cardiovascular diseases. The risk of acquiring many other different disorders can now also be reduced considerably by including regular physical exercise and change in eating habits.

It does not take sophisticated statistics to prove beyond a doubt that prevention is, in general, far more effective in keeping a person in good health than treatment. Many studies have shown evidence that a healthy lifestyle, eating well and not in excess, and a regular physical exercise regimen are a valid form of preventive action that can reduce the risk of acquiring many diseases. There is also evidence that many forms of supplements, such as vitamins and Omega 3 fatty acids, can improve the health of many people [149]. The full benefits of these readily available and inexpensive means are yet to be fully understood. Additionally, there are difficulties to get people to accept taking preventive action against diseases, such as consuming supplements, when a person has yet to become sick.

Another obstacle for getting people to accept the general use of preventive means to reduce the risk of diseases is related to the fact that only the likelihood of not getting a disease can be affected by prevention. Unfortunately, many people will wait to take action until they get the disease in question, and then they can only hope there is an effective cure available. The fact that general preventions do not carry any reward is an obstacle in itself. Prevention is by far the gentlest method of health care with minimal side effects and is difficult to bring forward because many people have an optimistic perception and expectation of medical and surgical treatment before they have experienced any treatment. People can also falsely claim that proper prevention cannot benefit everybody, and there will always be people who get sick despite taking all known preventive measures.

It, therefore, seems wise to shift the emphasis of some aspects of modern health care from the treatment of common diseases, such as medical and surgical procedures, to preventative care. This is primarily because prevention is far more effective and far less expensive than the treatments for many common diseases.

The use of preventive means to reduce the risk of getting diseases of various kinds has reduced health care costs substantially, and the full use of available preventive means would carry those savings further ahead and increase the living standard of many people.

People may have many reasons, valid and invalid, not to take appropriate actions in response to fear. People may falsely claim that they will not get the disease for which they have been vaccinated, so the vaccination would have been unnecessary in the first place. The problem is that nobody knows who or when they are going to catch a disease or have an accident. People do not plan to fall in the bathroom or catch a disease. Most people claim that they will not fall because "they are careful".

This statement has no scientific or medical value. Also, many people claim that they are better drivers of automobiles than the average person. Therefore, they believe they won't have an accident and do not need to wear seatbelts. The fact of the matter is people do indeed have accidents in doing very routine daily work, which they may assume to be perfectly safe.

Risks in medical and surgical treatments

Many people embark on medical and surgical treatments they later regret. Their regret stems from either the anticipated benefits did not materialize or because the side effects caused more harm than they had envisioned. If a person chose to have a surgical treatment that caused severe harm, the decision cannot be reversed. As a comparison, there is little, if any, side effects of preventive care.

One would think a rational person would evaluate the risks and benefits based on the likelihood that the anticipated benefit would be achieved, the risk of an unwanted event would be low, and the possible consequences minimal. However, most people handle health risks in utterly unsophisticated ways compared with methods used in business and finance, such as controlling economic risks. Most people still handle the risks to our health by going off of a "gut feeling". Similarly, decisions are made based on unsupported and sometimes biased information rather than quantitative measurements of risks. For example, people will accept screening test results or checkups without asking about the benefits. Additionally, only a few people ask about the drawbacks and complications.

When people do ask questions, people usually ask if a preventative method will help them to avoid illnesses or increase their life expectancy. People also accept surgical operations without knowing what risks are involved and, in particular, what these risks mean.

Worries about getting sick

Reducing the risk of getting sick can also minimize a person's worries about getting sick, thus reducing fear and anxiety. This is an advantage because of the availability of preventive measures, but they have received insufficient attention. Before the vaccine for poliomyelitis or polio was available, many people were scared of getting that devastating disease. The availability of an effective vaccine reduced the fear and anxiety of getting poliomyelitis for many individuals.

Concerns about getting sick can have a beneficial effect if it promotes taking appropriate actions that can reduce the risk of illness. However, concerns about getting sick where there are no known ways to reduce the risk of a particular illness have only adverse effects on a person.

There are now many different means available that can reduce the risk of getting various kinds of diseases, and those means can reduce the fear and anxiety that many people experience regarding illness and disease. The new ways of reducing the risk of diseases have spared many people from sickness and death, and other forms of treatment are continuously added to the arsenal of medical and surgical professionals. Also, new ways of treating diseases have improved the outcome of treatments of many diseases, and that has likewise reduced the fear and anxiety of getting many kinds of diseases.

The availability of effective treatments can also reduce the fear and anxiety of dying from becoming getting ill. Fortunately, the side effects of many kinds of treatments have been reduced as well. For example, surgical side effects of neurological deficits have been reduced by the use of a technique developed in the 1980s known as intraoperative neurophysiological monitoring [230].

The risk of acquiring infectious diseases, cardiovascular diseases, diabetes, and many other severe diseases can now be much reduced by simple means. Mothers can now reduce the risk of giving birth to babies with specific deficits such as spina bifida, autism, and several more disorders by taking a specific vitamin, folic acid, before and during their pregnancy. Unfortunately, many of the available preventive remedies are poorly utilized despite having a high degree of efficacy in reducing the risk of debilitating diseases, having a low risk of complications, and often being inexpensive. This is in contrast to the treatment of diseases that are often expensive and have severe and debilitating side effects such as encephalitis.

Physical exercise, lowering one's body weight, and refraining from alcohol and caffeine are effective ways to control blood pressure. The changing of one's lifestyle, including adding or increasing physical exercise, is an effective method in lowering blood pressure. Worrying and stress are prevalent causes of high blood pressure. Reducing stress and worry is also an efficient remedy for lowering blood pressure. Ultimately, some diseases, such as hypertension, can be treated effectively by medication, but pharmaceuticals often have negative side effects, while changing to a healthier lifestyle does not.

The fear of side effects from treatments or medications might be beneficial for healthcare, in general, because that fear may encourage preventive means. In the arena of healthcare, people's reaction to a certain treatment seem, to a great extent, to be controlled by what people believe rather than reliable facts. The term healthcare appears to be unique to the USA. In the US, the word healthcare means "care of the sick", while the name used in many other languages correctly translates to "sick care".

Studies have shown that it is difficult to change what people believe, and many people believe many things without asking or searching for evidence about the likelihood that what they worry about will even occur [17].

Prevention or treatment?

A person may face the task of deciding between the prevention of diseases or relying on treatment in the case of illness. Prevention and treatment have fundamental differences; prevention minimizes the risk of getting diseases, and treatment aims to restore normal function to the part or system of the body that is affected by the disease in question. Most preventive means have little or no side effects, and preventive means are usually inexpensive. In comparison, most forms of treatment have more severe side effects of various degrees. Also, many diseases lack efficient treatments.

Treatments are usually significantly more expensive than preventions. Preventions, on the other hand, have no immediate reward, and the rationale of taking action without any symptoms or signs of disease is disputed by many people. Preventive means, except vaccinations to some extent, are less used than what would be expected based on their efficacy and low rate of side effects.

There are now many preventative treatments available that can reduce the risk of acquiring many kinds of diseases. People, in general, must become more informed on the preventative treatments available to them. Methods of prevention, or rather minimizing of the risk of infectious diseases, include vaccinations. Additionally, there is more and more evidence from scientific studies that emphasize that some forms of healthy food, taking appropriate supplements, engaging in physical exercise, not smoking cigarettes, and refraining from drinking alcohol can reduce the risk of many severe diseases.

Vaccinations

Vaccinations have reduced the risk of getting any of many different infectious diseases to the extent of almost eliminating various diseases. Vaccinations for many infectious diseases do not only protect the person who is vaccinated, but also protects members of a community who may not be vaccinated.

This effect is caused by the fact that an epidemic cannot be sustained unless a certain fraction of the population cannot get the disease. This means that the decision of getting vaccinated cannot be left to an individual. To protect the population adequately, vaccinations must be controlled by laws and regulations. There are movements of people who, for one reason or another, are against vaccinations. Unfortunately, these movements are focus on false information or on vaccinations that were previously common childhood infections, such as measles.

Childhood vaccinations save millions of children from death and the crippling side effects caused by infectious diseases such as encephalitis, which is connected to the whooping cough disease (pertussis). In many states, children have to be vaccinated to get admission into the school. When it recently became easier to get an exemption for the childhood vaccination for the measles and whooping cough, cases of these diseases appeared again in the USA after having been nearly absent for many years.

In the year 2018, 80,000 people died in the influence in the USA. Sadly, very few of those people were vaccinated. It was claimed that the vaccine was not very effective, and it was reported that only 60% of individuals who were vaccinated escaped the symptoms of influenza. The fact that few of the people who died had been vaccinated means that the vaccine was effective in preventing death. Most people who die from influenza, in fact, die from pneumonia.

There is a very effective vaccine against pneumococcus pneumonia, which is effective for many years. If people took that in addition to the influenza vaccination, the combination would provide strong protection from death due to influenza. However, the pneumonia vaccine is only recommended for people over the age of 65 years when the risk of pneumonia increases. Young people also have a risk of dying from pneumonia, so it seems wise for young people to also receive the pneumonia vaccination, but the vaccination is not used to the extent of its efficacy.

In most states in the USA, it is by law required that children are vaccinated for several infectious childhood diseases such as measles, German measles (rubella virus), pertussis (whooping cough), and homophiles influenza. However, some physicians have decided to write letters in exception of the law for parents who wanted to avoid having their children vaccinated. This has created the basis for an epidemic. Continued outbreaks of pertussis, measles, and *H. influenzae* type b indicate that U.S. vaccination levels are inadequate in general.

Despite the phenomenal success of childhood vaccination, thousands of U.S. parents refuse specific vaccines or delay their administration. Some choose not to vaccinate their children at all. These parents are not a homogeneous group because some object to immunization on religious or philosophical grounds, some are avoiding a painful assault on their child, and others believe that the benefits of at least some immunizations do not justify the risks.

Since parents today have little or no experience with vaccine-preventable diseases, such as polio, *Hemophilus influenzae* type b, or measles, so they cannot easily appreciate the benefits of vaccination, the risks of not being vaccinated themselves, or the risk of not having their children vaccinated [231].

Throughout the recent measles outbreak in the USA, mothers who were against vaccinating their children claimed "medical freedom" as a justification for their refusal to have their children vaccinated. Sometime after the first vaccine against smallpox was developed in 1796, a similar antivaccination movement developed. At that time, the matter of allowing exemptions to the law of mandatory vaccinations ended up in the Supreme Courts of the USA.

"Jacobson v. Massachusetts, 197 U.S. 11 (1905), was a <u>United States Supreme</u> <u>Court</u> case in which the [Supreme] Court upheld the authority of the states to enforce compulsory vaccination laws. The Court's decision articulated the view that the freedom of the individual must sometimes be subordinated to the common welfare and is subject to the police power of the state."

The Supreme Court stated in its ruling: "The Fourteenth Amendment was brought up during the case on individual freedom. The case showed that a State was "restricting one aspect of liberty" by forcing people to get vaccinated. In its ruling in support of the Massachusetts law, the Supreme Court identified two primary rationales.

One was that "the state may be justified in restricting individual liberty... under the pressure of great dangers" to the safety of the "general public." By identifying the ongoing smallpox epidemic as a danger to the general public, the court ruled that individual rights and liberty were subordinate to the state's obligation to eradicate the disease. Jacobson had also argued that the law requiring vaccination was "arbitrary or oppressive." The Court rejected the argument stating that mandatory immunization in the face of an epidemic was not arbitrary or oppressive, but a measure for "getting to their goal of eradicating smallpox." Massachusetts was one of only 11 states that had compulsory vaccination laws.

Food supplements can diminish the risk of diseases

There is now considerable scientific evidence that the risk of severe diseases can be reduced by adding supplements to one's daily diet [149]. This study found evidence that taking 3,000 or 4,000IU of vitamin D_3 every day would reduce the risk of common diseases such as many forms of cancer. Many people now take supplements such as Omega 3 fatty acids, various kinds of vitamins, antioxidants, and other compounds.

The American College of Sports Medicine, the American Society of Clinical Oncology, the National Comprehensive Cancer Network, the American Cancer Society, the Oncology Nursing Society, the Commission on Cancer, and the Cancer Foundation now recommend physical exercise as an important part of the treatment for cancer. Also, many neurologists recommend physical exercise [232] for the treatment of neurological diseases such as moderate to severe depression. Keeping a normal body weight also plays an important role.

Recent studies have found great benefits from taking supplements such as Omega 3 [233]. In Germany, other studies showed that the risk of getting cancer could be reduced by as much as 50% if people would take 4,000 IUs of vitamin D_3 daily [148, 149]. It has been estimated that insufficient exposure to sunlight may kill up to 45,000 people in the USA each year from different internal organ cancers due to the lack of vitamin D_3 [234life, 235]. It has now become evident that the disadvantages of using sunscreen overshadow the advantages and that the general use of sunscreen has taken the lives of many people.

There are two reasons for taking vitamins as a supplement. One is to avoid the symptoms of deficits in essential substances such as vitamins, and the other reason is to achieve an optimal condition of body functions to promote a reduction in the risk of acquiring diseases of various kinds.

These two purposes of taking vitamins require very different vitamin levels in the blood. Achieving the optimal effects of vitamins, therefore, requires an intake of higher dosages of vitamins than what is required for avoiding deficit symptoms. Regular intake of food may be sufficient to achieve the blood levels required for avoiding signs of deficits, but the blood levels required for optimal benefits of these vitamins cannot be achieved from the regular intake of food. This can only be obtained by the intake of vitamins as food supplements.

Toxicity related side effects of vitamin D_3 occurs when blood levels are larger than 500nmol/L or equivalent to an intake of more than 500µg/L corresponding to 20,000IU [236]. The contribution of vitamin D_3 in normal food is small. Even milk, which is fortified with vitamin D_3, has only a small amount. A regular glass of milk contains about 50-100IU.

This means that the risk of acquiring many different diseases can now be reduced by making positive changes in one's lifestyle, eating habits, and food supplement regiment. Recently, the beneficial effect of supplements, especially vitamin D_3, and a healthy lifestyle has been supported by several studies [148] [149, 233].

The vast difference in the recommendation levels of vitamin consumption that is related to avoiding symptoms and signs from a deficit and the recommended levels for achieving the optimal advantages from the intake of vitamins has caused considerable confusion. RDA gives the amount that is assumed to be sufficient to avoid signs and symptoms of specific deficits, but the amounts required to achieve optimal organ functions are much higher. The dosage for avoiding symptoms of deficits is insufficient to provide the general advantages of intake of Vitamin D_3. However, the daily intake of vitamin D_3 that is necessary for getting the optional beneficial effect of vitamin D_3 is 4,000 IU [236].

As previously stated, a study in Germany concluded that if only one action was available to increase people's general health, it would be to consume 4,000 IU of vitamin D_3 every day [149]. Based on scientific studies, it has been estimated that the cost-saving effect of improving vitamin D_3 consumption in Germany might be as much as 37.5 billion Euros annually [149]. Also, the deficits of vitamin D_3 may be exaggerated and worsened by the use of sunscreen to avoid skin cancers, which are easier to treat and rarely causes deaths.

Preventions in the form of changes to one's lifestyle, including taking a supplement such the Omega 3 fatty acids and regularly doing physical exercise, have reduced the risk of acquiring many common diseases, such as various cardiovascular diseases and diabetes type 2. Omega-3 fatty acids are also used to treat hyperlipidemia and hypertension. It also can be used to reduce the risk of inflammatory diseases [237).

Also, Omega 3 can reduce high blood levels of triglycerides, {Skulas-Ray, 2008 #4101, 238] and it is effective in the treatment of inflammatory diseases such as arthritis and inflammatory bowel diseases [237] [239] [240]. It also has beneficial effects on the central nervous system in general. Studies have found it has a neuroprotective effect that may be beneficial in treating the effects of ischemic stroke [199, 241]. The beneficial effect due to intake of Omega 3 has been described and recommended for the treatment of mild to moderately severe depression [242].

However, there are no significant drug interactions with Omega-3 fatty acids. Higher dosages of Omega-3 fatty acids are required to reduce elevated triglyceride levels (2-4 g/day). Also, Omega 3 has an antiarrhythmic effect in addition to its antithrombotic properties [233].

The risk of cancer is on the mind of many people, and these worries cause much fear and anxiety. Not all of the risk factors of cancer are known, but some known risk factors can now be reduced. One of the effective means that can reduce the risk of getting cancer, but is lacking in effective treatment, is vitamin D_3. For example, the risk of colorectal cancer is inversely related to circulating vitamin D_3 [243].

In contrast, a recent published study claims to show that the intake of vitamin D_3 does not affect the risk of cancers. The study reported the results of a randomized, placebo-controlled trial, with a two-by-two factorial design, of vitamin D_3 (cholecalciferol) at a dose of 2,000 IU per day, [244] and marine n-3 (Omega-3) fatty acids at a dose of 1 g per day showed no reduction in the cancers in people over 50 years of age during the median follow-up time of 5.3 years [245]. This study, however, had too short of a duration to reveal any effect on cancer of the intake of vitamin D_3 administered the way it was administered in the study [244]. It takes years for most cancerous tumors to grow to the size where it is detectable with available clinical tests.

It is assumed that malignancies start with the mutation in one cell. This cell then divides into two cells, and then these two cells divide into four cells and so on. For breast cancer, the time for each division has been estimated to between one and two months. It may, therefore, take five years or more for a tumor to grow to a size that makes it detectable [246]. For lung cancers, it is estimated that it takes three to six months for a tumor to double in size. The news media published these results as an indication that supplements of vital D_3 and Omega-3 were useless in preventing or reducing the likelihood of getting cancer and cardiovascular diseases.

The fact that the study [245] covered only 5.3 years signifies that it did not adequately cover the time it takes for most tumors to develop to a size that can be detected by the available methods. If administered on the first day of the study and the intake of vitamin D_3 and Omega 3 prevented the mutations that would cause cancer, then the results would not become detectable with clinical diagnostic methods during the 5.3-year time of the study because it takes longer than five years for cancer to develop.

The intake of vitamin B_{12} is beneficial because it facilitates oxygen-rich blood reaching every cell in the body and helps boost nitric oxide and energy levels. Vitamin C is another antioxidant that helps fight free radical damage. Also, folic acid is an essential molecule that has many functions in the central nervous system. For example, folic acid helps the body make healthy new cells, as discussed in Chapter 4. Unfortunately, deficits in vitamin B and folic acid are present in many people. The impact of folic acid on the risk of the development of neuropsychiatric disorders in older individuals was reviewed in a recent article showing that the prevalence of folate deficiency is higher among individuals over the age of 65 mainly due to reduced dietary intake and intestinal malabsorption [247].

Studies have shown that taking folic acid can benefit cognitive functions in general [248]. Additionally, folic acid is important for the normal functioning of the central nervous system. It also has special functions related to reducing the risk of giving birth to a child with congenital malformations such as spina bifida, autism spectrum disorder, and more.

About forty-five years ago, it was shown that the administration of folic acid to women before and during pregnancy could substantially reduce the risk of giving birth to a child with spina bifida [249].

Various studies give different degrees of efficacy for this precaution of 70% to nearly 100% for spina bifida. The Centers for Disease Control and Prevention: Spina bifida cases declined up to 70% in 2002, and anencephaly cases declined up to 50% in 2003. The high reduction in the number of children with spina bifida after the introduction of folic acid before and during pregnancy, indicate that the condition can be eradicated if all women would take a folic acid tablet daily before and during pregnancy.

Spina bifida is one of the most common congenital disabilities and occurs in about 1.4 people per 1,000 pregnancies. It is a developmental disorder that causes severe disabilities because some of the spinal vertebrae are not fully formed before birth. After birth, the spinal vertebrae are not correctly connected and open. If the opening is large, a portion of the spinal cord can protrude through the opening. Unfortunately, surgery cannot sufficiently repair the defects, and people with spina bifida have severe symptoms such as leg weakness, hip dislocation, scoliosis, bladder and bowel problems, urinary tract infections, and a reduced kidney function.

Similarly, taking folic acid before and during pregnancy can also reduce the risk of giving birth to children with cleft palate [250]. A study in Mexico [251] found a 50% reduction in the incidence of anencephaly and spina bifida cases from 1999 to 2003 ($p<0.001$) from the administration of folic acid before and during pregnancy. Later it was shown that folic acid could also diminish the risk of other severe diseases such as autism [252, 253].

Surprisingly, such simple and inexpensive means that have no known side effects are not more widely used [253]. Now, more than 40 years after the finding that taking folic acid could reduce the risks of giving birth to a child with spina bifida was published, a study showed that only 35 % of mothers had taken folic acid before and during pregnancy [254]. That means that the real problem is in convincing people of the importance of these simple precautions. People happily do many checkups without asking if they are beneficial or not, but taking folic acid is not among what people decide to take. Could a woman's gynecologist find a more critical task than tell their patients to take folic acid before getting pregnant?

Influenza vaccination saves an average of 20,000 people from death every year in the USA. Last year, a reported 80,000 people died from influenza and its complications, and many more would be saved from the inconvenience of being sick.

The compliance with preventions, such as vaccination, is low in general. A recent study showed that only 30% of people receive influenza vaccines. Making vaccinations mandatory is challenging to get through the legislative process because it is said to infringe on personal freedom. It is more important to have effective laws that protect children because they cannot make decisions when it comes to actions that may have importance and influence on their entire life. Compliance with taking folic acid before and during pregnancy is equally low, at approximately 35% [254]

These are just a few of many examples where commonly available inexpensive preventive measures are ignored. One of the reasons many people do not use preventive means to reduce the risk of getting ill is that people are afraid of the side effects of preventing means. Another reason is that many people believe that prevention is ineffective. Other people believe that they will not get ill no matter what, and that it is only other people who get the diseases and conditions in question.

Many people do not achieve the full gain from many other known ways to reduce harm in general. For example, it has been necessary to create laws for getting people to comply with the use of seatbelts in personal cars. Currently, airlines refuse to let people fly without using seat belts, and children are not allowed to attend school without certain vaccinations.

Risk management

To briefly reiterate, the occurrence of most events is uncertain. Only the path of celestial bodies, such as the sun and the moon, can be predicted accurately. The occurrence of rain and other weather phenomena are not certain. Only the average rainfall after a storm is knowable. The occurrence of accidents and diseases are also uncertain, and it is not possible to know who will have an accident, when it will occur, who will become ill, when it will happen, and more. Although a person's death is certain, the time and date of when it will occur is uncertain. Only the likelihood of the occurrence of such events may be known. The likelihood of something happening can be affected however.

For example, drivers of automobiles have various variables and factors such as behaviors, skill levels, experience levels, road choices, time of travel, and several other factors that affect the risk of having a collision.

While the likelihood of an accident can often be determined accurately, it cannot be determined if and when a person will have one. This means decisions must be made under uncertainty. Risk management is a discipline for living with the possibility that future events may cause adverse effects [14]. Risk management makes it possible to increase the likelihood of a beneficial event and decrease the risk of harm. However, decisions based on information about the likelihood that something good or bad may occur are decisions made under uncertainty.

Risk management is the process of identifying, assessing, and controlling threats of various kinds. Risk management is used in many different disciplines, such as economics, business, psychology, engineering, and wherever decisions have to be made under uncertainty. Risk management is not widely used for decision making in most medical fields.

Decisions about health matters, whether using preventative treatments or waiting until the disease has manifested to take action, are not made using modern risk management methods. Unfortunately, the decision of which treatment to use when a person gets the symptoms of a disease is not made with the aid of modern risk management methods.

For example, the risks of having an accident when driving or riding in an automobile are not the same for all people in all possible situations. The risks are greater when the driver is drunk, which then has a great impact and affect to the risk of a motor vehicle accident (MVA). Therefore, deciding not to drive while drunk or deciding not to ride with a drunk driver is an effective way of reducing this specific risk. Selecting a skilled driver with a history of good judgment is another way to reduce the chance part of the risk equation.

Making decisions under uncertainties -similarities with gambling

The risks and benefits from medical and surgical treatment have many similarities with gambling. Some are winners and some are losers. Winning by gambling on roulette, rolling the dice, or on slot machines is entirely a matter of chance. The outcome of this kind of gambling is truly random. Generally, if repeated enough times, the average results can be calculated accurately. There are no decisions one can make that can affect the outcome. The casinos have adjusted the winning and losing on games like roulette and slot machines so that someone can periodically win or gain money. No rational person would pursue that kind of gambling for a long time because it is certain they will lose money.

Horse races, poker, and the stock market are examples where action based on knowledge can affect the chances of winning. This kind of gambling has two parts: a part that depends entirely on chance and a part that depends on skill and knowledge. Those who play these types of games attempt to minimize the part that depends on chance and have the part that depends on skill dominate. Many people are attracted to gambling, especially when gambling with a lot of money. History reveals that gambling was also common in ancient times as it is today. However, gambling with money is addictive and has led to much harm and despair for many people.

Many forms of medical and surgical treatments are forms of gambling that have the benefits or detriments as the outcome. This kind of gambling belongs to the kind of gambling where decisions by the patient play an important role in the outcome. Like other forms of gambling, making a choice of medical or surgical treatments also has two parts. The part that is pure chance and the part that depends on knowledge and skill.

A patients' knowledge about the risks is incomplete, as it is in many common diseases such as, for example, common back pain. Surgeons will often recommend an operation, but such treatment carries immediate risks of infections, death, and other various forms of motor or sensory function loss after a surgical operation. The likelihood of these risks and the type of adverse side effects varies from very benign and simple to death. Additionally, the risk of deficits and chronic pain after some operations is considerable [255].

For example, medical treatment using a non-steroidal anti-inflammatory (NSAID) is an alternative to surgical operations for the treatment of common lower back pain which is effective and does not have the risk of deficits and chronic pain. The negative symptoms of medical treatment for common lower back pain are related to the fact that the treatment requires a rigorous program of physical exercise. This means that the patient must have a considerable discipline over a long time in the administration of the medical treatment and engaging in physical exercise.

Many patients who have had surgery wish they had chosen medical treatment instead. Medical treatment can be stopped at any time if the patient changes his or her mind and it has no lasting effect on the patient's body. However, surgical treatment usually cannot be reversed after the operation has been performed. While much is known about the risk of unsuccessful surgical operations for common back pain [255], patients are not always informed about the risks and benefits of the different treatments.

There are many other situations in our daily lives where decisions must be made under uncertainty. For example, wearing a seatbelt when riding in a personal automobile is an example of decision made under uncertainty. A person who is traveling in an automobile not wearing a seat belt is gambling that an accident will not occur. Losing this game means that there is a high likelihood of death or disability. Winning means the person has the freedom to do as the person wants and the price for a more comfortable ride while not wearing a seatbelt is the risk of severe injury. The downside is considerably more severe.

Studies by economists have shown a person's perception of losing and winning the same amount of money, such as investing in the stock market or playing poker, is different. If these two feelings are compared, the bad feeling of losing a certain sum of money is perceived to be more extreme than the pleasure of gaining the same amount[256, 257].

This kind of reasoning is also valid in connection with making choices between different options in other areas of life where losing and winning are compared. It also applies to comparing the benefits of medical and surgical treatments with the harm that can occur.

In an ideal world, the evaluation of risks would be based on knowledge about what kind of harm might occur, how often this harm occurs, and other factors. What is claimed as bad luck, often is just the chance or uncertainty part of the risk equation has occurred unexpectedly. There is no such thing as good or bad luck, and Lady Luck is an illusion used in the place of understanding probabilities. A rational person would pay more attention to risks that have a high likelihood of occurring than the risks with a very low likelihood of occurrence. Likewise, a rational person would pay more attention to the magnitude of the risks that have serious consequences, such as death, over risks with less serious ones.

The real world is very different from person to person, and people often react irrationally when it comes to making decisions under uncertainty. One reason is that the outrage factor can increase or decrease people's perception of risk. One can say the outrage factor has a negative effect on most people's perception of healthcare. The consequences that people underestimate are the risks of injury and death in connection with medical or surgical treatments. (The outrage factor is described in Chapter 1).

Management of risks in health care

Making decisions about a person's health is not trivial; it is a form of making decisions under uncertainties. Decisions, such as choosing between medical and surgical treatment, are a kind of gambling where two factors are essential. One factor is the risk of severe complications, and the other factor is the degree of benefit from the treatment in restoring bodily functions to their value before the illness that caused treatment. This type of decision making is similar to gambling on the stock market where the choice is between an investment with a high outcome but a high risk of a catastrophic outcome, or a choice of an investment with much less risk but less to gain as well. In surgical and medical treatment, the choice is also often between a high-risk option of treatment where the outcome is excellent, but the risk of failure or complications is often high, and the other choice is a low risk option where the outcome may be less satisfactory yet has less risks of failure and complications. In some situations, there is a third option of choosing no treatment which may have advantages with less risk of adverse effects.

The treatment of a disease is not always beneficial to the patient, and it is important to keep in mind that treatment should not be done just because it is possible. Many disorders are grossly over-treated, while other disorders are not treated adequately or accurately.

Psychological harm is usually not considered as serious as bodily harm, but it can no doubt have a negative impact on an individual's quality of life. Since most forms of medical and surgical treatment involve the risk of harm, avoiding treatment sometimes means less harm.

A person's choice of treatment often depends on the person's risk tolerance. This would be similar to the choices a person makes when investing, such as in the stock market. When gambling on the stock market, a person can very much choose the level of risk. Different people chose different levels of risk, mainly because different people have different levels of risk tolerance. The choice depends on a person's tolerance to risk-taking behavior. This has similarities with gambling on the stock market because the results depend on reliable and detailed information about the stocks as well as the choice of risk level a person is prepared to take. Taking high risks are associated with high returns, and therefore, the choice of what kind of stocks a person wants to invest in depends on a person's risk tolerance. The main difference between making choices regarding the management of a person's health and investing in the stock market is people have a poorer basis of relevant data for healthcare related choices than the data available to a stock trader.

Many physicians and surgeons in the USA have two roles. The first role is providing advice to the patient about what to do, and the second is to conduct the treatment. In addition, many physicians and surgeons have an economic interest in what they advise the patient about. A surgeon is more likely to recommend surgery, and a physician, such as an internist, neurologist, or cardiologist, is more likely to recommend treatment with medications. Also, a surgeon's advice depends on the surgeon's specialty. A more complex treatment, such as a surgical operation, provides a substantially higher income for the surgeon than treatment with medications and follow-up counseling regarding change of lifestyle. This may affect the advice a patient receives. The two roles of being an advisor and a provider of a service, therefore, may have the risk of being in conflict with each other.

In the ideal world, the patient will regard the physician's suggestion of treatment as just that. It is solely a suggestion. The patient should then ask for a second opinion and possibly do some research of their own. Then the patient will have a more thorough understanding of the risks and benefits of the different treatments available, and then a patient can select to have the treatment or no treatment at all that gives the greatest benefit with the least risk of harm.

In the real world, patients will understand only a part of what the physician tells them about the cure rate (benefits) and risk of side effects, but some of the information will be forgotten or misinterpreted by the patient. If the patient is not used to making decisions under uncertainty, then the patient, therefore, cannot make up his/her mind about what treatment is best.

Instead, the patient will probably take the physician's recommendation without requiring further explanations, either because of difficulties in deciding which treatment to choose or because he/she trusts the authority of the physician.

The fact that physicians have a dual role, in which they are a source of advice about treatment but are also the ones who receive the economic benefits, may be the cause of some of these problems. Similar double roles have been identified in other parts of our society, such as in the administration of assets to individuals or groups of people.

The meaning of the word "need."

Selecting a patient's treatment is often complicated and often requires considerable information and elaborations. Naturally, it is easier to state that a person needs to have a specific treatment, such as an operation or a specific medication. The use of the word "need" is convenient because it eliminates the necessity to answer other questions from a patient, such as what benefit will the treatment provide, what are the risks, or what are the harmful side effects. Rarely is it asked what "need" means, and often it is not made clear why the treatment is necessary or "a need". Is the treatment "necessary" because it is beneficial for maintaining life itself or is it beneficial in preserving the quality of life that was enjoyed before becoming ill? Alternatively, if the treatment is "necessary," is it assumed that the treatments will prevent death or give the patient a more enjoyable life? What is the likelihood that the goals of a specific treatment will have the beneficial results that the physician described to the patient?

Uncertainty can cause a great amount of fear and anxiety; thus, it is important that patients know what a physician or surgeon means when using the words he/she uses in communicating with patients.

Harm and errors from medical tests and treatment

Diagnostic tests

Diagnoses are not always correct, but are often believed by patients anyhow. People place high confidence in diagnostic tests, but like any other assessment that is based on measurements and human interpretation, the results or diagnosis can be made in error by the physician. Also, diagnostic tests have errors and uncertainties, and their interpretation may be incorrect, thus causing incorrect diagnoses.

That uncertainty can be decreased by having a second opinion regarding a diagnosis, but the effect of an error cannot be reduced by doing more of the same kind of tests. Only by using different equipment and different people to interpret the results can errors be reduced. Scientists always repeat tests that give unusual or abnormal values, but this is not common in connection with diagnostic tests.

An incorrect diagnosis can cause two kinds of harm. One kind is bodily harm caused by being treated for a disease in which the patient does not have. The other kind of harm from an incorrect diagnosis is related to the fear and anxiety that an incorrect diagnosis can cause and the subsequent treatment that results in no relief of the patient's symptoms.

Additionally, abnormal test results may cause fear in the patient, and it would be in the patient's interest to have a second opinion as soon as possible. Knowing a diagnosis that indicates a person has a disease causes great amounts of fear. Many kinds of tests involve risks, such as from exposure to ionizing radiation, infection risks from invasive tests, or even a needle biopsy or insertions of a catheter in one of the openings of the body can result in an infection. If a procedure is done in a hospital, the infection can be severe because of hospital bacteria may be involved. Also, many tests are done that are not beneficial to a patient, and many tests are done without a clear purpose or goal.

There was a time when a patient's diagnosis was kept a secret from the patient, and that was followed by a time where healthcare personnel took pride in being open and telling patients everything regarding their diagnosis including preliminary results and interpretations that were uncertain. In fact, interpreting test results always has some degree of uncertainty. That change to transparency is supposed to be beneficial for the patients, but when a preliminary diagnosis is given to a patient, it is more likely to cause more harm than good. It can also be confusing when the final diagnosis is different from the preliminary diagnosis.

Routine checkups

Routine checkups are sometimes regarded to be preventive means for diseases. At best, routine checkups can detect diseases earlier than when the patient would show signs of symptoms. Checkups only provide a history of the state of a person's body, not what will happen after that the checkup was done. Many diseases are resolved without treatments, and many patients recover spontaneously without any treatment. There are even some cases where tumors stop growing or even decrease in size and a few vanish without treatment [258].

The purpose of using extensive routine checkups for detecting diseases that have not yet produced symptoms is related to the assumption that treatment is more effective if started early in the disease. This has strong commonsense support, but it is not valid for all diseases. There are diseases that either are not life-threatening, such as some forms of prostate cancer that grows so slowly that it will not kill a person before other diseases do, or where treatment cannot change the disease's progression. Routine screenings for people who do not have any symptoms or signs of disease and who do not have the diseases in question in their family will have the risk of false positive test results. These false results may lead to treatment for a disease that the person does not have and a considerable amount of distress which can cause anxiety.

In a report published 2014, The Swiss Medical Board (www.medical-board.ch) found that systematic mammography screening might prevent about one breast cancer related death for every 1000 women screened [259]. The report also found no evidence that suggested that the overall mortality rate was affected by mammograms. Also, the report emphasized the importance of the harm associated with conventional mammograms and, in particular, it pointed to the false positive test results which contributed to the risk of over-diagnosis.

In a 2011 study by Wilkinson et al., it was found that screening using mammography led to a 30% increase in over-diagnosis and overtreatment, or an absolute risk increase of 0.5 % [260]. This means that some healthy women, who would not have been diagnosed if there had not been a screening, will be treated unnecessarily [260]).

These authors also reported that for every breast cancer death prevented in women on a 10-year regiment of annual screenings beginning at 50 years of age in the U.S., 490 to 670 women are likely to have a false positive mammogram with repeat examination. Of these biopsies performed, as many as 70 to 100 were not beneficial to the patients, and 3 to 14 were over-diagnosed with breast cancer, which would never have become clinically apparent [261]. These findings resulted in the revised recommendation of the use of mammograms [259].

The risk of inducing cancer by the radiation used in mammography, estimated to be one in 1,000 for a lifetime of mammography, may now be reduced because of new methods used in mammography [259, 262]. Also, the effect of a person's mental well-being is rarely taken into consideration when the value of mammography and other screenings, such as checkups, are discussed.

Most of the women who have a mammogram have increased anxiety before the examination is done, as well as when the women receive the results of the test. Women who have routine mammograms and have no symptoms or a hereditary predisposition for breast cancer have shown that negative test results often elicit anxiety about what the next test might show.

It is easy to claim that nobody dies from anxiety, but it certainly decreases a person's quality of life. In fact, there is evidence that anxiety may increase the risk of diseases related to its increase in the activity of the sympathetic nervous system.

Test of blood pressure and blood sugar (and A1C)

The test for blood pressure and blood sugar are different from other forms of routine checkups. They are different because the results of these tests show signs that indicate diseases may occur in the future.

Prediabetes and the early stages of diabetes have no symptoms, but having prediabetes and the early stages of diabetes may lead to actual diabetes and high risks of damage to the kidney and other organs [263] [264]. Changes in diet, increased physical exercise, and losing weight if overweight or obese are effective in reversing prediabetic signs. Thus, changing one's lifestyle can make it possible to avoid diseases.

If the early prediabetic signs are detected and remedied early, diabetes can be avoided. Early detection of elevated blood sugar and HbA1C levels and appropriate intervention can reduce the risks of organ damage and other severe diseases. Also, hypertension is an important risk factor for stroke and cardiovascular diseases. Signs of these diseases, such as elevated blood sugar (or A1C, glycated hemoglobin) and high blood pressure, can be detected in routine checkups.

High blood pressure has been known for many years to be risk factors for severe diseases such as cardiovascular diseases [265] and stroke [266]. However, it has now become evident that high blood pressure is also a risk factor for neurological diseases such as memory loss and possible dementias of various kinds including Alzheimer's disease. Furthermore, the recommendation for treatment of hypertension before it does damage has been lowered, and it is now recommended that resting systolic blood pressure should not be higher than 120. Additionally, it may be beneficial to take actions to lower systolic blood pressure that is higher than 120. High blood pressure can be reversed by physical exercise and control of body weight. If exercise and a healthy diet is not sufficient, then there are effective medications that can control blood pressure.

Risk of harm from common medications

Adverse effects of acetaminophen in adults

The intake of acetaminophen (common brand name is Tylenol) affects many body systems and the consequences can be severe. Acetaminophen is toxic to the liver in dosages above 4,000 mg per 24 hours. Recent studies have brought a new understanding of the side effects of acetaminophen. In particular, the studies are concerned with the oxidative metabolism of acetaminophen, and many known and several new acetaminophen-metabolites have been discovered from these studies [148]. Studies have also indicated immune-modulating effects at a 2,000 mg dose and oxidative stress responses at a 4,000 mg dose [267].

There are also indications that the intake of acetaminophen may affect brain development. This is especially important since it is now the most widely used medication in children. Acetaminophen is readily available for infants in an over-the-counter form labeled as "safe, gentle, and effective," with no warnings of side effects other than allergic reactions [268]. Despite compelling evidence that acetaminophen has many severe side effects, including liver failure in adults, it is still advertised as having fewer side effects than other nonsteroidal anti-inflammatory pain relievers.

In the USA, a recent study showed that acetaminophen intake was the most frequent cause of acute liver failure. Acetaminophen accounts for 46% of acute liver failure, followed by undetermined causes (15%) and idiosyncratic drug reactions [269]. This means that most of the acute liver failures had preventable causes.

Of the acute liver failure patients, 26% required emergency liver transplants. Acetaminophen-induced liver failure was the cause of the majority of patients who had these liver transplants. A study concluded that acetaminophen liver toxicity far exceeds other causes of acute liver failure in the United States [228].

Acetaminophen has a profound effect on adult brain function, such as blunting the response to both negative and positive stimuli. This includes a blunted response to threatening stimuli and a reduction in behavioral responses to social rejection [268]. Also, there is evidence that taking acetaminophen impairs the ability of most adults to identify errors made during the performance of simple or common tasks. Many people believe that acetaminophen is a safe analgesic and ignore the side effects. That belief may be the reason for its extensive use, not only as a primary analgesic, but also as an addition to many other medications, including sleep medications and anti-allergy medications that can be bought without prescriptions.

The general belief that acetaminophen is a safe medication that is effective for controlling acute pain is another example of where beliefs and facts collide, causing a false feeling of security.

Prenatal effect of acetaminophen

Recently, studies have found that the intake of several commonly used substances can increase the risk of congenital disabilities. The adverse effects of a commonly used painkiller, acetaminophen (Tylenol), or paracetamol was first published by Schulz et al. 2008 [270]. These results were later confirmed and extended by other investigators. Other investigators, such as Stergiakouli in 2016 [271], reported an association between the intake of acetaminophen during pregnancy and emotional symptoms of various kinds. Also, the effect of acetaminophen is different for children in the age group of birth to 4 or 5 years of age, older children, and adults. Some of these severe effects that were described in the scientific literature were explained recently in a report in "ProPublica" [272]. The report from 2013, highlights the side effects of Tylenol and other products containing acetaminophen [148]. The report also declared Tylenol or acetaminophen to be the deadliest over-the-counter medication on the U.S. market. Few people seem to know that a relatively small overdose of acetaminophen can be deadly.

Published in 2011, the results of a multinational study of more than 200,000 children showed that the use of acetaminophen might increase the risk of developing and maintaining asthma by as much as 40% [273]. Other investigators have noted that the marked increase in autism, asthma, and ADHD that began in the early 1980s was coincident with the time that aspirin was replaced by acetaminophen. In 2014, Thompson et al. showed that earlier claims that findings of increased ADHD with acetaminophen use are specific to acetaminophen [274]. These investigators found that even low doses of acetaminophen can affect behavior seven years later. This is alarming because acetaminophen is the most commonly used antenatal or prenatal drugs.

Different groups of investigators [272] have concluded that children exposed to acetaminophen prenatally are at increased risk of multiple behavioral difficulties. Also, these associations do not appear to be explained by unmeasured behavioral or social factors insofar as these symptoms are not observed for postnatal or partner's acetaminophen use. The advertising of Tylenol and other acetaminophen products are ignoring these side effects, thus an example of commercial interests over the welfare of people.

Adverse effects of acetaminophen in 0-5 year children

The most recent revelations of the severe side effects from the intake of acetaminophen concern acetaminophen's effect on young children. In 2008, Schultz and co-workers found that acetaminophen was significantly associated with autism in children of the age of five or less. However, ibuprofen use after measles-mumps-rubella vaccination was not associated with autistic disorders [270]. Another cause that may have been involved in triggering the autism epidemic is an artificial sweetener named aspartame [275]. The evidence against aspartame is weaker than that for acetaminophen [148].

The complex action and metabolism of acetaminophen in babies and children are shown in Figure 6.1 [148].

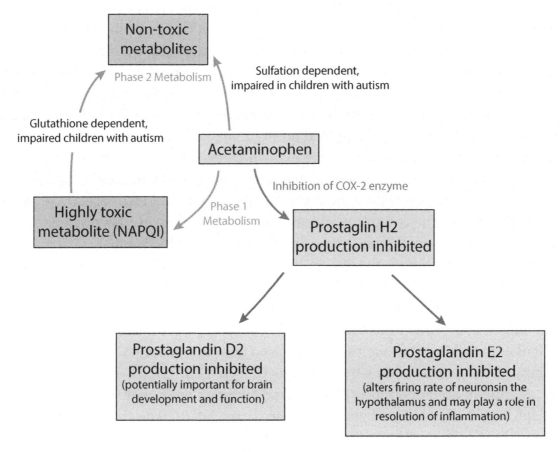

Figure 6.1 Action and metabolism of acetaminophen in babies and children. Based on Parker, 2017) [148] (Artwork by Liliana Cabrera.)

These recent studies have confirmed that giving young children acetaminophen increases the risk of autism spectrum diseases [148, 276], and that oxidative stress and inflammation promotes the development of similar conditions such as autism. Recent studies have shown that the risk of acquiring other diseases seems to increase in the presence of oxidative stress. Oxidative stress, like inflammation, is associated with an increased risk of cancer, coronary artery disease, and many other psychiatric disorders. It is widely thought that inflammation and oxidative stress go hand in hand in increasing the risk of severe diseases [148].

A model of increasing the risk of autism is depicted in Figure 6.2 and is based on the findings that inflammatory reactions depend on inflammatory mediators which are the biome depletion, chronic psychological stress, vitamin D deficiency, inflammatory diets, and a sedentary lifestyle. The risk factors for this kind of inflammation is increased with the age of the parents at the time of birth, autoimmune conditions, obesity, and diabetes.

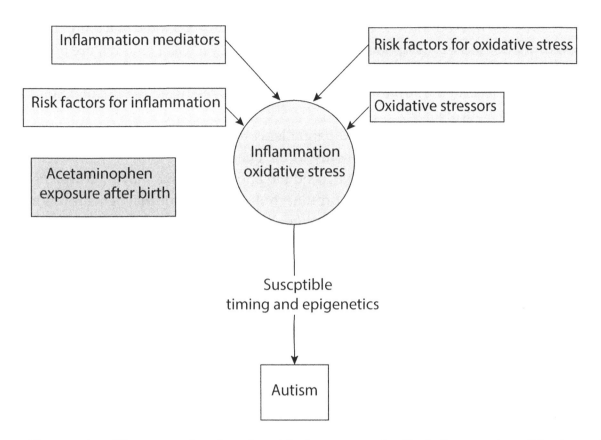

Figure 6.2. The central role of oxidative stress (red), inflammation (blue), and possibly exposure to acetaminophen after birth in the induction of autism. (Based on Parker, 2017)[148]. (Artwork by Liliana Cabrera.)

The oxidative stressors that are involved are infections, exposure to heavy metals, organophosphates, cigarette smoke, and air pollution. This also includes methanol, jaundice, and the problems associated with vitamin B metabolism.

Adverse effects of opioids

There are two severe adverse effects of the use of opioids in the treatment and management of pain. One severe effect is the risk of addiction, and the other risk is the suppression of breathing when consumption exceeds a certain amount. Also, there are many side effects of opioids in general. Itching may occur as a prominent side effect due to the administration of opioids [277]. However, this form of itching is not caused by the release of histamine. An antagonist to opioids known as Naloxone can inhibit mu and kappa opioid receptors and reduce side effects of opioids [278] [279].

For pain control, the most commonly used forms of opioids are semi-synthetic opioids such as oxycodone (OxyContin, Percocet), hydrocodone (Vicodin, Lortab), hydromorphone, and oxymorphone. Synthetic opioids, such as Fentanyl, are generally used for treating many forms of acute pain. However, the recent crisis regarding the addictive side effects of the administration of opioids has restricted the use of opioids for the treatment of acute pain.

The excessive prescribing of opioid painkillers by physicians where NSAIDs would have been sufficient is contributing greatly to the overuse. It is generally believed that the semi-synthetic opioids, such as hydrocodone and oxycodone, have a lower risk of addiction than other opioids, such as morphine, but solid data for that belief is lacking.

There is some evidence that street drugs, such as heroin, have a higher risk of addiction compared to the overused semi-synthetic opioids. However, the evidence is, in general, weak and inconclusive. Three earlier studies showed little risk of side effects in treatment using opioids except for addiction, but it has been concluded that the basis for these conclusions was weak [280-282].

The original meaning of the term "addiction" when in reference to a drug means that withdrawal causes distinct symptoms. The fact that all opioids are treated as having a similar ability to cause addiction is also a problem. Heroin and the family of semi-synthetic opioids are different in their ability to cause addiction, but people in general, including journalists, do not focus on this major difference.

Opioids are often prescribed for treating chronic neuropathic pain for which they have limited efficacy in eliminating a person's pain. Unfortunately, many patients will increase the dosage of opioids they take when what was prescribed does not relieve their pain. This can be a route to addiction, and it can cause death because overdosage of opioids can cause respiratory arrest. If a person cannot get a sufficient supply of opioids from his or her physician, then he or she may look for a street vendor to buy opioid pain killers. The vendor may offer heroin as a less expensive source.

The fact that opioids as pain relievers become ineffective after approximately 12 weeks of use because of one's tolerance to frequently prescribed drugs such as oxycodone or hydromorphone is another problem. This problem can lead people increasing their dosage of opioids, even though the effect of opioids wanes because of the tolerance factor [283] [284].

The mood-related effect of opioids in creating a sense of well-being has greatly complicated the evaluation of the effect of long-term use of opioids in the treatment of chronic pain. The rewarding effect of opioids and the analgesic potency are related to the state of mood they create, and this sense of well-being may somehow affect the pain by creating an indifference to the pain [285].

Adverse effects of consumptions of alcohol

The intake of alcohol has many different effects. Alcohol is addictive. Thus, alcohol has an effect on social relationships. Alcohol alters people's judgment, which makes people make bad decisions that they would not normally make when sober, such as driving a motor vehicle while intoxicated or having risky sexual intercourse. The lack of good judgment may increase the risk of injury or having accidents such as motor vehicle accidents. Alcohol may change the expressions of a person's personality. Alcohol consumption has profound effects on many bodily functions, such as the absorption of vitamins. The intake of high quantities of alcohol affects many bodily functions such as motor functions, and it has adverse effects on the liver and other body organs. Ethanol-induced oxidative stress plays a significant role in the mechanisms by which ethanol produces liver injury. Alcohol consumption, especially in great amounts, is a strong risk factor for liver diseases [286, 287].

Heavy alcohol consumption, four or more drinks/day, is significantly associated with a 5-fold increased risk for oral cancer, pharyngeal cancer, and esophageal squamous cell carcinoma, a 2.5-fold for laryngeal cancer, a 50% increase for colorectal and breast cancers, and a 30% increase for pancreatic cancer. Alcohol's risk factor of cancer has previously been the object of large epidemiological studies [288, 289]. Thus, from these studies, there is little evidence of a threshold effect [288], which means that the consumption of any amount of alcohol, large or small, will increase the risk of many diseases.

Preventable death and injuries from accidents

Overconfidence makes people conclude that it is not necessary to take precautions, such as not wearing seat belts or driving when under the influence of alcohol, tranquilizers, or narcotics. Overconfidence regarding a person's driving skills is one of the reasons that as many as 80% of all drivers claim that they are better drivers than the average driver (Kahneman et al. p469-470) [256]. This claim represents a noticeable overestimation of driving skills. The situation and perspective for a person changes drastically the day an accident happens to him or her, but is it necessary to have an accident to realize that accidents indeed do occur to the average person? One would hope that more people would learn from other's mistakes. For example, the risk of many kinds of accidents that often happen at home could be reduced if people were aware that falling is a severe risk, particularly for the elderly.

Simple precautions, such as not walking up/down the stairs when intoxicated, not being in a hurry when walking or running, using proper footwear, and taking precautions when walking outside when the ground is icy, could reduce the number of severe head injuries, various broken bones, or broken hips that often lead to death in the elderly. Also, the use of proper footwear and not being in a hurry when walking at home or anywhere else could considerably reduce the risk of falling in general. The floor in the bathroom is often slippery after becoming wet. Simply enough, placing a towel on the floor of the shower reduces the risk of falling. Additionally, educational programs regarding accidents that teach people not to drive when intoxicated or to use a safety belt whenever driving or riding in an automobile might be the most beneficial way to get more people to take preventive measures.

The total number of "preventable" deaths in the United States in 2017 was 169,936 people and injuries were about 47.2 million. Thus, the estimated economic costs from these unintentional deaths and injuries were $1,034.6 billion. Preventable injuries are the third leading cause of death in the USA year 2016. (Data from "Injury Facts" from The National Safety Council). Unintentional injury and unintentional death are terms that are used by the National Safety Council in connection with accidents.

Surprisingly, most people who die from preventable injuries are at home, and 72% of fatal injuries occur at home or in public with almost all of them being preventable (National Safety Council, 2019). This means that there is a great amount of room for reductions in the risks of such unintentional death.

The lifetime odds of death for selected causes in the United States, 2016:

Heart disease and cancer: 1 in 6
Motor vehicle crashes: 1 in 102
Opioid pain killers: 1 in 109
Falls: 1 in 119
Gun assaults: 1 in 285
Drowning: 1 in 1,086
Fire or smoke: 1 in 1,506
Choking on Food: 1 in 3,138
Sunstroke: 1 in 8,976
Lightning: 1 in 114,195.
(Data from "Injury Facts" The National Safety Council, nsc.org., 2018.)

Effect of alcohol intoxication on the risk of traffic accidents

Every day, almost 30 people in the United States die in drunk-driving crashes. That is one person every 48 minutes in 2017. These deaths have fallen by a third in the last three decades. However, drunk-driving accidents still claim more than 10,000 lives per year. In 2010, the most recent year for which cost data is available, these deaths and damages resulted in a cost of $44 billion for that year (Data from NHTSA.gov, 2018).

It is mainly the poor judgments made by a driver that makes intoxicated driving dangerous. The fact that drivers who are under the influence of alcohol have impaired judgment is supported by the interesting, but also frightening, observation that only 16% of the intoxicated drivers with a BAC of 0.1 % or more who died in MVAs wore safety belts (26.4 % for those with BAC less than 0.1 %). Alternatively, 41.5 % of sober drivers wore safety belts. (Safety Facts, National Safety Council, 1995, p87).

Why do people drive while intoxicated? Naturally, there are always people who claim that the previous instances where they drove while intoxicated should serve as proof that it is not dangerous to drive while intoxicated. Again, this is similar to playing Russian Roulette. It may go well for someone many times in a row, but if one plays the game a sufficient number of times, they will lose.

Abbreviations

A

ACC, anterior cingular cortex
ACh acetylcholine
ACG anterior cingulate gyrus
acetylcholinesterase
ACTH adrenocorticotropic hormone
AH anterior hypothalamus
ALA Alpha-linolenic acid
AMG amygdala
Amy amygdala
ANS, autonomic nervous system
AP area postrema
ATP, adenosine triphosphate

B

BA basal amygdala
BA Brodman classification of
cerebral cortex
BD bipolar disorder
BDNF brain-derived neurotrophic
factor
BL Basolateral nucleus of the
amygdala
BLA basolateral amygdala
BM basomedial nucleus
BNST bed nucleus of stria terminalis
BOLD Blood-Oxygen-Level
Dependent

C

CA cornu ammonis
CA1 cornu ammonis 1
CA3 cornu ammonis 3
CBD Cannabidiol
CeA Central nucleus, amygdala
CEm medial subdivision, amygdala

CNX vagus nerve
CR conditional response
CRH, corticotropin-releasing
hormone
CS conditional stimulus

D

DA dopamine
dACC dorsal anterior cingulate
cortex
DC dorsal cortex (of the IC)
DCM dynamic causal modeling
DG dentate gyrus
DLPFC dorsolateral prefrontal
cortex,
DMN dorsal motor nucleus vagus
nerve
DPT diphtheria-tetanus-pertussis
DSM-5 Diagnostic and Statistical
Manuel
DTI diffusion tensor imaging

E

E2 Estradiol
EBL emotional body language
EEG electroencephalographic
EMG electromyography
EMS emotional motor system

F

FDA Food and Drug
Administration
FFG fusiform gyrus
FG fusiform gyrus
FM fibromyalgia

fMRI functional magnetic resonance imaging

FPS fear-potentiated startle

FSH follicle-stimulating hormone

G

GABA, γ-aminobutyric acid

GAD generalized anxiety disorders

GAD67 glutamic acid decarboxylase

GI tract

Glu glutamate

GR glucocorticoid receptor

H

HCS healthy control subjects

HF hippocampal formation

Hipp hippocampus

HP, hippocampus

HPA hypothalamus-pituitary-adrenal

HPG hypothalamic–pituitary–gonadal

HYP hypothalamus

I

IC inferior colliculus

ICX external inferior colliculus

IL infralimbic (structures of the amygdala)

IL interleukin

IL-1, IL-6, IL-8 pro-inflammatory cytokines,

IL-1β interleukin-1β

IL-6 interleukin-6

Ins insula

IOM Institute of Medicine

IPS intraparietal sulcus

ITC intercalated cells

K

kD kilodalton

L

LA Lateral nucleus (of the amygdala)

LC locus coeruleus

LD Lateral dorsal nucleus

LDT lateral dorsal tegmentum

LGN lateral geniculate nucleus

LH lateral hypothalamus

LH luteinizing hormone

LHb lateral habenula

lITC lateral intercalated cell cluster

LOC lateral occipital complex

LP: Lateral posterior nucleus

LPS lipopolysaccharide (endotoxin)

LS lateral septum

M

mAChR muscarinic acetylcholine receptor

MAPK mitogen-activated protein kinase

MD mediodorsal thalamus

MDD major depressive disorder

MdPFC, dorsomedial prefrontal cortex

MeA medial amygdala

MEG magnetoencephalographic

MGB medial geniculate body

mITC medial intercalated cell cluster

mPFC medial prefrontal cortex

MPFC medial prefrontal cortex (),

MPOA the medial preoptic area of the hypothalamus

MSN medium spiny neurons

MTL medial temporal lobe
MTL, medial temporal lobe
MVA motor vehicle accidents

N

NA nucleus ambiguous
NAc nucleus accumbens
NE norepinephrine
NeP neuropathic pain
NLRs nucleotide-binding
oligomerization domain-like receptors
NO nitrogen oxide
NSA National Security Agency
NST nucleus of the solitary tract
NTS nucleus tractus solitarius

O

OB, olfactory bulb
Occ cort occipital cortex
OFC orbital prefrontal cortex
orbFC orbitofrontal cortex
OT oxytocin OB olfactory bulb

P

PAG periaqueductal grey
PBR parabrachial region
PCC posterior cingulate cortex
PET positron emission tomography
PFC prefrontal cortex
PGs prostaglandins
PI3K phosphoinositide 3-kinase
PIN posterior intralaminar nucleus
Pit, pituitary gland
PL prelimbic (structures of the
amygdala)
PM premotor cortex
PTSD post-traumatic stress disorder
Pulv pulvinar thalamus

PVN paraventricular nucleus of
hypothalamus

R

RDA recommended daily allowance
RDI recommended daily intake
RTMg, rostromedial tegmentum
RVLM rostroventrolateral medulla
RVM rostroventral medulla

S

SAD social anxiety disorders
SC superior colliculus
SDQ Strengths and Difficulties
Questionnaire
SG suprageniculate nucleus
SON supraoptic nucleus
SP substance P
SS somatosensory
SSRI selective serotonin reuptake
inhibitors
STD sexually transmitted diseases
STS superior temporal sulcus
STT spinothalamic tract

T

THC Tetrahydrocannabinol
TLR4 toll-like receptor
TNF-α tumor necrosis factor-α
TRN thalamic reticular nucleus

U

UF uncinate fasciculus

V

VA Ventral anterior nucleus

vACC ventral anterior cingulate cortex

vHC ventral hippocampus

VI ventral intermediate nucleus

VL ventral lateral nucleus

VLPFC ventrolateral prefrontal cortex

VLPFC, ventrolateral prefrontal cortex

VMH ventromedial hypothalamus

VMM ventromedial medulla

vmPFC ventromedial prefrontal cortex

VP ventral pallidum

VP ventral posterior nucleus

VPM ventral posteriomedial nucleus

VPN ventral posterolateral nucleus

VS ventral striatum

VTA ventral tegmental area

References

1.   Møller, A.R., *Sensory Systems: Anatomy and Physiology*. 2014, Aage R. Møller Publishing: Dallas.

2.   Young, W.R. and A.M. Williams, *How fear of falling can increase fall-risk in older adults: applying psychological theory to practical observations*. Gait Posture, 2015. **41**(1): p. 7-12.

3.   Pegna, A.J., et al., *Discriminating emotional faces without primary visual cortices involves the right amygdala*. Nat. Neurosci., 2005. **8**: p. 24-25.

4.   Jiang, Y. and S. He, *Cortical responses to invisible faces: dissociating subsystems for facial-information processing*. Curr. Biol., 2006. **16**: p. 2023–2029.

5.   Adolphs, R., *The Biology of Fear*. Current Biology, 2013. **23**: p. R79-R93.

6.   Garrido, M.I., et al., *Functional evidence for a dual route to amygdala*. Curr. Biol., 2012. **22**: p. 129–134.

7.   Tamietto, M., et al., *Subcortical connections to human amygdala and changes following destruction of the visual cortex*. Curr. Biol., 2012. **22**: p. 1449–1455.

8.   Pessoa, L., R. Adolphs, and N.N. Rev., *Emotion processing and the amygdala: from a "low road" to "many roads" of evaluating biological significance*. Nat. Neurosci. Rev., 2010. **11**: p. 773–782.

9.   Berridge, K.C. and P. Winkielman, *What is an unconscious emotion? (The case for unconscious "liking")*. Cognition Emotion, 2003. **17**: p. 181–211.

10. Krogsbøll, L., K. Jørgensen, and P. Gøtzsche, *General health checks in adults for reducing morbidity and mortality from disease*. JAMA, 2013. **309**(13): p. 2489-90.

11. Wakefield, A., et al., *Ileal-lymphoid-nodular hyperplasia, non-specific colitis, and pervasive developmental disorder in children*. Lancet., 1998. **351**(9103).

12. Deer, B., *How the vaccine crisis was meant to make money*. BMJ, 2011. **342**:(5258).

13. Sandman, P.M., *Risk communications: Facing public outrage*. Environmental Protection Agency (EPA) Journal, Washington, 1987: p. 21-22.

14. Bernstein, P.L., *Against the Gods. The remarkable story of risk*. 1996, New York: John Wiley and Sons.

15. Wheelock, M., et al., *Threat-related learning relies on distinct dorsal prefrontal cortex network connectivity*. Neuroimage, 2014. **102**: p. 904-12.

16. O'Donovan, A., et al., *Exaggerated neurobiological sensitivity to threat as a mechanism linking anxiety with increased risk for diseases of aging*. Neurosci Biobehav Rev. , 2013. **37**(1): p. 96-108.

17. Gorman, S.E. and J. Gorman, *Denying to the Grave: Why We Ignore the Facts That Will Save Us*. 2016.

18. Hertz, N., *Eyes Wide Open: How to Make Smart Decisions in a Confusing World*. 2013: Harper Collins.

19. Berns, G., et al., *Neurobiological Correlates of Social Conformity and Independence During Mental Rotation*. Biol Psychiatry, 2005. **58**(3): p. 245-53.

20. Garakani, A., S. Mathew, and D. Charney, *Neurobiology of anxiety disorders and implications for treatment*. Mt Sinai J Med, 2006. **73**(7): p. 941-9.

21. Öhman, A., *Fear and anxiety: Evolutionary, cognitive, and clinical perspectives*, in *Handbook of emotions.*, M.L.J.M. Haviland-Jones, Editor. 2000, The Guilford Press.: New York. p. 573–593.

22. Dias, B., S. Banerjee, and K. Ressler, *Towards new approaches to disorders of fear and anxiety.* Current opinion in neurobiology, 2013. **23**(3): p. 346-352.

23. Salmon, P., *Effects of physical excercise on anxiety, depression, and sensitivity to stress: A unifying theory.* Clinical Psychology Review, 2001. **21**(1): p. 33-61.

24. Parsons, R.G. and K.J. Ressler, *Implications of memory modulation for post-traumatic stress and fear disorders.* Nature Neuroscience, 2013. **16**: p. 146–153.

25. Pawluski, J., J. Lonstein, and A. Fleming, *The Neurobiology of Postpartum Anxiety and Depression.* Trends in Neurosciences, 2017. **40**(2): p. 106-120.

26. Malizia, A., et al., *Decreased brain GABA(A)-benzodiazepine receptor binding in panic disorder: preliminary results from a quantitative PET study.* Psychiatry, 1998. **55**: p. 715–720.

27. Lorberbaum, J., et al., *Neural correlates of speech anticipatory anxiety in generalized social phobia.* Neuroreport., 2004. **15**: p. 2701–2705.

28. Clark, D.M. and A. Wells, *A cognitive model of social phobia,* in *Social Phobia: diagnosis, assessment, and treatment,* R. Heimberg, et al., Editors. 1995, Guilford: New York. p. 69–93.

29. Nagata, T., F. Suzuki, and A. Teo, *Generalized social anxiety disorder: A still-neglected anxiety disorder 3 decadessince Liebowitz's review.* Psychiatry Clin Neurosci, 2015. **69**(12): p. 724-40.

30. Griffin, J., et al., *The effect of self-reported and observed job conditions on depression and anxiety symptoms: a comparison of theoretical models.* J Occup Health Psychol., 2007. **12**(4): p. 334-49.

31. Mathew, S., J. Coplan, and J. Gorman, *Neurobiological mechanisms of social anxiety disorder.* Am J Psychiatry, 2001. **158**(10): p. 1558-67.

32. Barrett, L., *Emotions are real.* Emotion, 2012. **12**(3): p.:413-29.

33. Folkman, S. and J.T. Moskowitz, *Coping: Pitfalls and Promise.* Annu. Rev. Psychol., 2004. **55**: p. 745-74.

34. Zeki, S. and J.P. Romaya, *Neural correlates of hate.* PLoS One, 2008. **3**(10): p.:e3556.

35. LeDoux, J.E., *Anxious: Using the Brain to Understand and Treat Fear and Anxxiety.* 2015, New York: Viking.

36. Silverstein, D.N. and M. Ingvar, *A multi-pathway hypothesis for human visual fear signaling.* Front Syst Neurosci., 2015. **9**(9): p. 101.

37. Gandhi, S.P., D.J. Heeger, and G.M. Boyton, *Spatial attention effects brain activity in human primary visual cortex.* Proc. of National Academy of Science, 1999. **96**(6): p. 3314-3319.

38. Møller, A.R., *New Developments in Neuroscience. A Review.* Archives of Neurology and Neurosurgery, 2019. **2**(1): p. 48-58.

39. Jaén, I., *The Romantic Syndrome: A Neuropsychological Perspective.* http://www.cognitivecircle.org/ct&lit/CogCircleResearch/CogLit_Emo.html.

40. LeDoux, J.E., *The emotional brain: The Mysterious Underpinnings of Emotional Life* Vol. Images. 1998: Simon & Schuster.

41. Karnath, H., O. and B. Baier, *Right insula for our sense of limb ownership and self-awareness of actions.* Brain Struct Funct., 2010. **214**(5-6): p. 411-7.

42.   Benes, F.M., *Amygdalocortical Circuitry in Schizophrenia: From Circuits to Molecules.* Neuropsychopharmacology, 2010. **35**(1): p. 239–257.

43.   Jacobs, B. and D. McGinty, *Participation of the amygdala in complex stimulus recognition and behavioral inhibition: evidence from unit studies.* Brain Res, 1972. **36**: p. 431–436.

44.   Paré, D. and G. Quirk, *When scientific paradigms lead to tunnel vision: lessons from the study of fear.* NPJ Science of Learning, 2017: p. 2-6.

45.   Blanchard, C., et al., *The brain decade in debate: III. Neurobiology of emotion.* Braz J Med Biol Res. , 2001. **34**(3): p. 283-93.

46.   Dagleish, T., *The emotional brain.* Nature Reviews Neuroscience, 2004. **5**: p. 583–58.

47.   Papez, J.W., *A proposed mechanism of emotion.* J Neuropsychiatry Clin Neurosci., 1937. **7**(1): p. 103-12.

48.   LeDoux, J., *Emotion, memory and the brain.* Sci Am., 1994. **270**(6): p. 50-7.

49.   LeDoux, J.E., *Emotional Networks in the Brain,* in *Handbook of Emotions.,* M. Lewis and J. Havilland, Editors. 1993, The Guilford Press: New York.

50.   Ehrlich, I., et al., *Amygdala inhibitory circuits and the control of fear memory.* Neuron, 2009. **62**(6): p. 757-7.

51.   Weinberger, N., *The medial geniculate, not the amygdala, as the root of auditory fear conditioning.* Hearing Research, 2010. **274**: p. 61-74.

52.   Do-Monte, F., K. Quiñones-Laracuente, and G. Quirk, *A temporal shift in the circuits mediating retrieval of fear memory.* Nature, 2015. **519**(7544).

53.   Sotres-Bayon, F. and G. Quirk, *Prefrontal control of fear: more than just extinction.* Current Opinion in Neurobiology, 2010. **20**(2): p. 231-235.

54.   Amir, A., et al., *Amygdala signaling during foraging in a hazardous environment.* J. Neurosci., 2015. **35**: p. 12994–13005.

55.   Brodal, P., *The Central Nervous System Fourth Edition.* 2010, New York: Oxford University Press.

56.   Downer, J.L.C., *Changes in visual agnostic functions and emotional behaviors following unilateral temporal pole damage in the "split-brain" monkey.* Nature, 1961. **191**: p. 50-51.

57.   Graeff, F.G., *Serotonin, the periaqueductal gray and panic.* Neurosci. Biobehav. Rev., 2004. **28**: p. 239–259.

58.   Klüver, H. and P.C. Bucy, *"Psychic blindness" and other symptoms following bilateral temporal lobectomy in rhesus monkeys.* Am. J. Physiol., 1937. **119**: p. 352-353.

59.   Arruda-Carvalho, M. and R.L. Clem, *Prefrontal-amygdala fear networks come into focus.* Front Syst Neurosci., 2015. **9**: p. 145.

60.   Münsterkötter, A., et al., *Spider or no spider? Neural correlates of sustained and phasic fear in spider phobia.* Depress Anxiety., 2015.

61.   Mobbs, D., et al., *Neural activity associated with monitoring the oscillating threat value of a tarantula.* Proc Natl Acad Sci U S A, 2010. **107**(47): p. 20582-6.

62.   Ohman, A., et al., *On the unconscious subcortical origin of human fear.* Physiol Behav, 2007. **92**(1-2): p. 180-5.

63.   Møller, A.R., et al., *Contribution from crossed and uncrossed brainstem structures to the brainstem auditory evoked potentials (BAEP): A study in human.* Laryngoscope, 1995. **105**: p. 596-605.

64.   Møller, A.R. and P. Rollins, *The non-classical auditory system is active in children but not in adults.* Neurosci. Lett., 2002. **319**: p. 41-44.

65.   Lockwood, A.H., *Tinnitus.* Neurologic clinics, 2005. **23**: p. 893-900.

66.   Møller, A.R., J.K. Kern, and B. Grannemann, *Are the non-classical auditory pathways involved in autism and PDD?* Neurol Res, 2005. **27**: p. 625-629.

67.   LeDoux, J.E., *Brain mechanisms of emotion and emotional learning.* Curr. Opin. Neurobiol., 1992. **2**: p. 191-197.

68.   Malmierca, M.S., et al., *A novel projection from dorsal cochlear nucleus to the medial division of the medial geniculate body of the rat.* Association for Research in Otolaryngology Abstracts, 2002. **25**: p. 176.

69.   Rudrauf, D., et al., *Rapid interactions between the ventral visual stream and emotion-related structures rely on a two-pathway architecture.* J Neurosci., 2008. **28**(11): p. 2793-803.

70.   Morrison, A.R., L.D. Sanford, and R.J. Ross, *The amygdala: a critical modulator of sensory influence on sleep.* Biological Signals & Receptors, 2000. **9**(6): p. 283-96.

71.   Morrison, D.A. and J. Sacks, *Balancing benefit against risk in the choice of therapy for coronary artery disease. Lesson from prospective, randomized, clinical trials of percutaneous coronary intervention and coronary artery bypass graft surgery.* Minerva Cardioangiol., 2003. **51**(5): p. 585-97.

72.   Møller, A.R., M.B. Møller, and M. Yokota, *Some forms of tinnitus may involve the extralemniscal auditory pathway.* Laryngoscope, 1992. **102**: p. 1165-1171.

73.   Lockwood, A., et al., *The functional neuroanatomy of tinnitus. Evidence for limbic system links and neural plasticity.* Neurology, 1998. **50**: p. 114-120.

74.   Gothard, K., et al., *Neural responses to facial expression and face identity in the monkey amygdala.* J Neurophysiol., 2007. **97**(2): p. 1671-83.

75.   Adolphs, R., et al., *Fear and the Human Amygdala.* Journal of Neuroscience, 1995. **15**(9): p. 5879-5891.

76.   Damasio, A.R., H. Damasio, and G.W. Van Hoesen, *Prosopagnosia: anatomic basis and behavioral mechanisms.* Neurology, 1982. **32**(4): p. 331-41.

77.   Carr, J., *I'll take the low road: the evolutionary underpinnings of visually triggered fea.* Front Neurosci, 2015. **9**: p. 414.

78.   Morris, J., et al., *A differential neural response in the human amygdala to fearful and happy facial expressions.* Nature, 1996. **383**(6603): p. 812-5.

79.   Kuraoka, K., N. Konoike, and K. Nakamura, *Functional differences in face processing between the amygdala and ventrolateral prefrontal cortex in monkeys.* Neuroscience, 2015. **304**: p. 71-80.

80.   Price, J.L., *Comparative aspects of amygdala connectivity.* Ann N Y Acad Sci, 2003. **985**: p. 50-8.

81.   McDonald, A.J., *Cortical pathways to the mammalian amygdala.* Progr. Neurobiol., 1998. **55**(3): p. 257-332.

82.   Møller, A.R., *New Developments in Neuroscience.* Journal of Integrated Creative Studies, 2015. **1**: p. 1-23.

83.   Marsh, R.A., et al., *A novel projection from the basolateral nucleus of the amygdala to the inferior colliculus in bats*. Soc. Neurosci. Abstr., 1999. **25**: p. 1417.

84.   Møller, A., *Neuroplasticity and Its Dark Sides: Disorders of the Nervous System, Second Edition*. 2nd ed. 2018, Dallas, Texas: Aage R. Møller. 244.

85.   Cahill, L., et al., *Is the amygdala a locus of "conditioned fear"? Some questions and caveats*. Neuron, 1999. **23**(2): p. 227-8.

86.   Olsson, A. and E. Phelps, *Social learning of fear*. Nature Neuroscience 2007. **10**(9): p. 1095–1102.

87.   Fast, C. and J. McGann, *Amygdalar Gating of Early Sensory Processing through Interactions with Locus Coeruleus*. J Neurosci, 2017. **37**(11): p. 3085-3101.

88.   Kilgard, M.P. and M.M. Merzenich, *Plasticity of temporal information processing in the primary auditory cortex*. Nature Neurosci., 1998. **1**: p. 727-731.

89.   Kilgard, M.P. and M.M. Merzenich, *Cortical map reorganization enabled by nucleus basalis activity*. Science, 1998. **279**: p. 1714-1718.

90.   Bakin, J.S. and N.M. Weinberger, *Induction of a physiological memory in the cerebral cortex by stimulation of the nucleus basalis*. Proc. Natl. Acad. Sci. USA, 1996. **93**(20): p. 11219-24.

91.   Weinberger, N.M., *Learning-induced physiological memory in adult primary auditory cortex: Receptive field plasticity, model, and mechanisms*. Audiol. Neuro-Otol., 1998. **3**: p. 145-167.

92.   Gloor, P., *The temporal lobe and limbic system*. 1997, New York: Oxford Press.

93.   Klüver, H. and P.C. Bucy, *Preliminary analysis of functions of the temporal lobes in monkeys*. Arch. Neurol. Psychiat., 1939. **42**: p. 979-1000.

94.   Adolphs, R., et al., *Impaired recognition of emotion in facial expressions following bilateral damage to the human amygdala*. Nature Neuroscience, 1994. **372**: p. 669-672.

95.   Terzian, H. and O. Dalle, *Syndrome of Kluver and Bucy reproduced in man by bilateral removal of the temporal lobes*. Neurology., 1955. **5**: p. 373–380.

96.   Ghika-Schmid, F., et al., *Klüver-Bucy syndrome after left anterior temporal resection*. Neuropsychologia, 1995. **33**: p. 101–113.

97.   Lilly, R., et al., *The human Klüver-Bucy syndrome*. Neurology., 1983. **33**(1141–1145).

98.   Hayman, L., et al., *Klüver-Bucy syndrome after bilateral selective damage of amygdala and its cortical connection*. J Neuropsychiatry Clin Neurosci, 1998. **10**(3): p. 354-8.

99.   Gloor, P., *Review on the amygdala*, in *Handbook of Physiology*, F. J., Editor. 1960, American Physiological Society. p. 1395–1420.

100.   Blanchard, D. and R. Blanchard, *Innate and conditioned reactions to threat in rats with amygdaloid lesions*. J. Comp. Physiol. Psychol, 1972. **81**: p. 281–290.

101.   Kellicut, M. and J. Schwartzbaum, *Formation of a conditioned emotional response (CER) following lesions of the amygdaloid complex in rats*. Psychol. Rev, 1963. **12**: p. 351–358.

102.   LeDoux, J.E., *Rethinking the emotional brain*. Neuron, 2012. **73**(4): p. 653-676.

103.   LeDoux, J.E., et al., *The lateral amygdaloid nucleus: Sensory interface of the amygdala in fear conditioning*. J. Neurosci., 1990. **10**: p. 1062-1069.

104.   LeDoux, J.E., A. Sakaguchi, and D.J. Reis, *Subcortical efferent projections of the medial geniculate mediate emotional responses conditioned by acoustic stimuli*. J. Neurosci., 1984. **4**: p. 683-698.

105. McIntyre, C.K., J.L. McGaugh, and C.L. Williams, *Interacting brain systems modulate memory consolidation.* Neurosci Biobehav Rev., 2011.

106. Ferry, B., B. Roozendaal, and J.L. McGaugh, *Role of norepinephrine in mediating stress hormone regulation of long term memory storage: a critical involvement of the amygdala.* Biol. Psychiatry, 1999. **46**(9): p. 1142-1152.

107. De Ridder, D., et al., *Amygdalohippocampal involvement in tinnitus and auditory memory.* Acta Otolaryngol, 2006. **Suppl(556)**: p. 50-53.

108. Neugebauer, V., et al., *The amygdala and persistent pain.* Neuroscientist, 2004. **10**(3): p. 221-34.

109. Møller, A.R., *Neurophysiologic abnormalities in autism*, in *New Autism Research Developments*, B.S. Mesmere, Editor. 2007, Nova Science Publishers: New York. p. 137-158.

110. Baron-Cohen, S., et al., *The amygdala theory of autism.* Neurosci Biobehav Rev, 2000. **24**(3): p. 355-64.

111. Halpern, M., *The organization and function of the vomeronasal system.* Ann. Rev. Neurosci., 1987. **10**: p. 325-62.

112. LeDoux, J.E., *The emotional brain.* 1996, New York: Touchstone. 384.

113. Doron, N.N. and J.E. LeDoux, *Cells in the posterior thalamus project to both amygdala and temporal cortex: a quantitative retrograde double-labeling study in the rat.* J. Comp. Neurol., 2000. **425**: p. 257-274.

114. LeDoux, J.E., *Synaptic self.* 2002, New York: Viking.

115. Schlee, W., et al., *Age-related changes in neural functional connectivity and its behavioral relevance.* BMC Neurosci., 2012. **13**(1).

116. Le, Q., et al., *Snakes elicit earlier, and monkey faces, later, gamma oscillations in macaque pulvinar neurons.* Sci Rep., 2016. **6**(20595).

117. Bechara, A. and N. Naqvi, *Listening to your heart: interoceptive awareness as a gateway to feeling.* Nature Neuroscience, 2004. **7**: p. 102 - 103.

118. Hermans, E., et al., *Fear bradycardia and activation of the human periaqueductal grey.* Neuroimage, 2013. **66**: p. 278-87.

119. Paulus, M. and M. Stein, *An insular view of anxiety.* Biol. Psychiatry, 2006. **60**: p. 383–387.

120. Damasio, A. and G. Carvalho, *The nature of feelings: evolutionary and neurobiological origins.* Nat Rev Neurosci., 2013. **14**(2): p. 143-52.

121. Stephani, C., et al. *Stimulation of the insula.* in *Second Congress, International Society of Intraoperative Neurophysiology.* 2009. Dubrovnik.

122. Stephani, C., et al., *Functional neuroanatomy of the insular lobe.* Brain Struct Funct., 2010

123. Seth, A., *Interoceptive inference, emotion, and the embodied sel.* Trends Cogn Sci., 2013. **17**(11): p. 565-73.

124. Strata, P., *The Emotional Cerebellum.* Cerebellum, 2015.

125. Sacchetti, B., B. Scelfo, and P. Strata, *Cerebellum and emotional behavior.* Neuroscience., 2009. **162**: p. 756-62.

126. Schutter, D. and J. van Honk, *The cerebellum on the rise in human emotion.* Cerebellum., 2005. **4**: p. 290–4.

127. Schmahmann, J.D. and J.C. Sherman, *The cerebellar cognitive affective syndrome.* Brain, 1998. **121**: p. 561–79.

128. Perciavalle, V., et al., *Consensus paper: current views on the role of cerebellar interpositus nucleus in movement control and emotion.* Cerebellum, 2013. **12**: p. 738–57.

129. Mahan, A. and K. Ressler, *Fear conditioning, synaptic plasticity and the amygdala: implications for posttraumatic stress disorder.* Trends Neurosci., 2012. **35**(1): p. 24-35.

130. Roozendaal, B. and J. McGaugh, *Memory modulation.* Behav Neurosci., 2011. **125**(6): p. 797-824.

131. Cahill, L., et al., *Beta-adrenergic activation and memory for emotional events.* Nature., 1994. **371**(6499): p. 702–704.

132. Yerkes, R.M. and J.D. Dodson, *The Relation of Strength of Stimulus to Rapidity of Habit-Formation.* Journal of Comparative Neurology and Psychology, 1908. **18**: p. 459-482.

133. De Quervain, D.J., B. Roozendaal, and J.L. McGaugh, *Stress and glucocorticoids impair retrieval of long-term spatial memory.* Nature, 1998. **394**(6695): p. 787-790.

134. Floor, H., *Maladaptive plasticity, memory for pain and phantom limb pain: Review and suggestions for new therapies.* Expert Rev Neurotherapeutics, 2008. **8**: p. 809-818.

135. Mahan, A.L. and K.J. Ressler, *Fear conditioning, synaptic plasticity and the amygdala: implications for posttraumatic stress disorder.* Trends Neurosci., 2012. **35**(1): p. 24-35.

136. LaBar, K.S. and R. Cabeza, *Cognitive Neuroscience of Emotional Memory.* Nature Reviews Neuroscience, 2006. **7**: p. 54-64.

137. Hermans, E., et al., *How the amygdala affects emotional memory by altering brain network properties.* Neurobiol Learn Mem., 2014. **112**: p. 2-16.

138. de Gelder, B., *Towards the neurobiology of emotional body language.* Nature Reviews Neuroscience, 2006. **7**: p. 242-249.

139. McGaugh, J.L., *Memory consolidation and the amygdala: a systems perspective.* Trends Neurosci, 2002. **25**(9): p. 456.

140. Mayer, E.A., et al., *Stress and irritable bowel syndrome.* American Journal of Physiology-Gastrointestinal and Liver Physiology, 2001. **280**(4): p. G519-G524.

141. Mayer, E.A., *The neurobiology of stress and gastrointestinal disease.* Gut, 2000. **47**: p. 861–869.

142. Weinberger, N.M., R. Javid, and B. Lepan, *Long-term retention of learning-induced receptive-field plasticity.* Proc. Nat. Acad. Sci, 1993. **90**: p. 2394-2398.

143. Mayer, E.A. and K. Tillisch, *The Brain-Gut Axis in Abdominal Pain Syndromes.* Annu Rev Med., 2011. **62**: p. 381-396.

144. Chekroud, A.M., et al., *A review of neuroimaging studies of race-related prejudice: does amygdala response reflect threat?* Front Hum Neurosci., 2014. **8**(179): p. 179.

145. Salzman, C.D. and S. Fusi, *Emotion, cognition, and mental state representation in amygdala and prefrontal cortex. .* Annu. Rev. Neurosci. , 2010. **33**: p. 173–202.

146. Maeng, L.Y. and M.R. Milad, *Sex differences in anxiety disorders: Interactions between fear, stress, and gonadal hormones.* Horm Behav., 2015. **76**: p. 106-17.

147. Panksepp, J., *Affective Neuroscience: The Foundations of Human and Animal Emotions.* 1998, New York: Oxford University Press.

148. Parker, W., et al., *The role of oxidative stress, inflammation and acetaminophen exposure from birth to early childhood in the induction of autism.* J Int Med Res., 2017. **45**(2): p. 407-438.

149. Zittermann, A., *The estimated benefits of vitamin D for Germany.* Mol Nutr Food Res., 2010. **54**(8): p. 1164-71.

150. Santaolalla, R. and M. Abreu, *Innate immunity in the small intestine.* Gastroenterology, 2012. **28**(2): p. 124–129.

151. Pavlov, V. and K. Tracey, *The vagus nerve and the inflammatory reflex—linking immunity and metabolism.* Nat Rev Endocrinol. , 2012. **8**(12): p. 743–754.

152. Bushnell, M.C., M. Ceko, and L. Low, *Cognitive and emotional control of pain and its disruption in chronic pain.* Nature Reviews Neuroscience, 2013. **14**(7): p. 502-511.

153. Phillips, M., C. Ladoucer, and W. Drevets, *A neural model of voluntary and automatic emotion regulation: implications for understanding the pathophysiology and neurodevelopment of bipolar disorder.* Mol Psychiatry, 2008. **13**: p. 833–857.

154. Tabbaa, M., Paedae, B, Liu, Y, Wang, Z, *Neuropeptide Regulation of Social Attachment: The Prairie Vole Model.* Compr Physiol. , 2016: p. 81-104.

155. Lonstein, J., et al., *Emotion and mood adaptations in the peripartum female:complementary contributions of GABA and oxytocin.* J Neuroendocrinol., 2014. **26**(10): p. 649-64.

156. Romano, A., et al., *From Autism to Eating Disorders and More: The Role of Oxytocin in Neuropsychiatric Disorders.* Front Neurosci, 2016. **9**: p. 497.

157. Craig, A., *How do you feel? Interoception: the sense of the physiological condition of the body.* Nat Rev Neurosci., 2002. **3**(8): p. 655-66.

158. Frith, U. and F. Happé, *Autism: beyond "theory of mind.* Cognition, 1994. **50**(1-3): p. 115-32.

159. Harlow, H.F. and R.R. Zimmermann, *Affectional responses in the infant monkey; orphaned baby monkeys develop a strong and persistent attachment to inanimate surrogate mothers.* Science, 1959. **130**: p. 421-32.

160. Meaney, M.J., *Maternal care, gene expression, and the transmission of individual differences in stress reactivity across generations.* Annu Rev Neurosci, 2001. **24**: p. 1161-1192.

161. Diorio, J. and M. Meaney, *Maternal programming of defensive responses through sustained effects on gene expression.* J Psychiatry Neurosci, 2007. **32**(4): p. 275-84.

162. Quattrocki, E.F., K., *Autism, oxytocin and interoception.* Neurosci Biobehav Rev., 2014. **47**: p. 410-430.

163. Russo, S.J. and E.J. Nestler, *The brain reward circuitry in mood disorders.* Nature Reviews Neuroscience 2013. **14**: p. 609–625

164. Lynch, J.F., et al., *Activation of ERβ modulates fear generalization through an effect on memory retrieval.* Horm Behav., 2014. **66**(2): p. 421-9.

165. Anonymous, *Risks and Benefits of Estrogen Plus Progestin in Healthy Postmenopausal Women.* JAMA, 2002. **288**(3): p. 321-333.

166. Glover, E.M., T. Jovanovic, and S.D. Norrholm, *Estrogen and extinction of fear memories: implications for posttraumatic stress disorder treatment.* Biol Psychiatry., 2015. **78**(3).

167. Cover, K.K., et al., *Mechanisms of estradiol in fear circuitry: implications for sex differences in psychopathology.* Transl Psychiatry, 2014. **4:e422**.

168. Blask, D., *Melatonin, sleep disturbance and cancer risk.* Sleep Med Rev, 2009. **13**(4): p. 257-64.

169. Rondanelli, M., et al., *Update on the role of melatonin in the prevention of cancer tumorigenesis and in the management of cancer correlates, such as sleep-wake and mood disturbances: review and remarks.* Aging Clin Exp Res., 2013. **25**(5): p. 499–510.

170. Ali, T. and M. Kim, *Melatonin ameliorates amyloid beta-induced memory deficits, tau hyperphosphorylation and neurodegeneration via PI3/Akt/GSk3β pathway in the mouse hippocampus.* J Pineal Res, 2015. **59**(1): p. 47-59.

171. Elliott, R., et al., *Affective cognition and its disruption in mood disorder.* Neuropsychopharmacology, 2011. **36**(1): p. 153-82.

172. Mayberg, H.S., *Targeted electrode-based modulation of neural circuits for depression.* J Clin Invest., 2009. **19**(4): p. 717-25.

173. Matthews, S., et al., *Decreased functional coupling of the amygdala and supragenual cingulate is related to increased depression in unmedicated individuals with current major depressive disorde.* J Affect Disorder, 2008. **111**: p. 13–20.

174. Maletic, V. and C. Raison, *Neurobiology of depression, fibromyalgia and neuropathic pain.* Front Biosci, 2009. **14**: p. 5291-5338.

175. Jänig, W. and H.J. Häbler, *Sympathetic nervous system: contribution to chronic pain.* Prog. Brain Res., 2000. **129**: p. 451-68.

176. Noesselt, T., et al., *Asymmetrical activation in the human brain during processing of fearful faces.* Curr Biol., 2005. **15**(5): p. 424-9.

177. Dejean, C., et al., *Neuronal Circuits for Fear Expression and Recovery: Recent Advances and Potential Therapeutic Strategies.* Biological Psychiatry, 2015. **78**(5): p. 298-306.

178. McGaugh, J., *Memory reconsolidation hypothesis revived but restrained: theoretical comment on Biedenkapp and Rudy* Behav Neurosci, 2004. **118**: p. 1140-2.

179. Wall, P.D., *The presence of ineffective synapses and circumstances which unmask them.* Phil. Trans. Royal Soc. (Lond.), 1977. **278**: p. 361-372.

180. Bilkei-Gorzo, A., et al., *Dynorphins regulate fear memory: from mice to men.* J Neurosci. , 2012. **32**(27).

181. Lonergan, M.H., et al., *Propranolol's effects on the consolidation and reconsolidation of long-term emotional memory in healthy participants: a meta-analysis.* J Psychiatry Neurosci. , 2013. **38**(4): p. 222–231.

182. Heaton, L.J., D.W. McNeil, and P. Milgrom, *Propranolol and d-cycloserine as adjunctive medications in reducing dental fear in sedation practice.* SAAD Dig, 2010. **26**: p. 27-35.

183. Fani, N., et al., *Fear-potentiated startle during extinction is associated with white matter microstructure and functional connectivity.* Cortex., 2015. **64**: p. 249-59.

184. Shin, L. and I. Liberzon, *The neurocircuitry of fear, stress, and anxiety disorders.* Neuropsychopharmacology, 2010. **35**: p. 169–191.

185. Tovote, P., J. Fadok, and A. Lüthi, *Neuronal circuits for fear and anxiety.* Nature Reviews Neuroscience, 2015. **16**(6): p. 317-331.

186. Duvarci, S. and D. Pare, *Amygdala microcircuits controlling learned fear.* Neuron, 2014. **82**(5): p. 966–980.

187.   Peters, J., P. Kalivas, and G. Quirk, *Extinction circuits for fear and addiction overlap in prefrontal cortex.* Learn Mem, 2009. **16**(5): p. 279–288.

188.   Maren, S., K. Phan, and I. Liberzon, *The contextual brain: implications for fear conditioning, extinction and psychopathology.* Nature Reviews, 2013. **14**: p. 417-428.

189.   Mochcovitch, M., et al., *A systematic review of fMRI studies in generalized anxiety disorder: evaluating its neural and cognitive basis.* J Affect Disord. , 2014. **167**: p. 336-42.

190.   Bartlett, A., R. Singh, and R. Hunter, *Anxiety and Epigenetics.* Adv Exp Med Biol, 2017. **978**: p. 145-166.

191.   Kessler, R.C., et al., *Prevalence, severity, and comorbidity of twelve-month DSM-IV disorders in the National Comorbidity Survey Replication (NCS-R).* Arch. Gen. Psychiatry, 2005. **62**: p. 617–627.

192.   Schiller, D., et al., *From fear to safety and back: reversal of fear in the human brain.* J. Neurosci., 2008. **28**: p. 11517–11525.

193.   Scharmüller, W., A. Wabnegger, and A. Schienle, *Functional Brain Connectivity During Fear of Pain: A Comparison Between Dental Phobics and Controls.* Brain Connect, 2014.

194.   Scharmüller, W., A. Wabnegger, and A. Schienle, *Functional Brain Connectivity During Fear of Pain: A Comparison Between Dental Phobics and Controls.* Brain Connectivity, 2015. **5**(3): p. 187-191.

195.   Sahoo, S., N. Hazari, and S. Padhy, *Choking Phobia : An Uncommon Phobic Disorder, Treated with Behavior Therapy : A Case Report and Review of the Literature.* Shanghai Arch Psychiatry, 2016. **28**(6): p. 349-352.

196.   Etkin, A. and T.D. Wager, *Functional neuroimaging of anxiety: a meta-analysis of emotional processing in PTSD, social anxiety disorder, and specific phobia.* Am J Psychiatry, 2007. **164**(10): p. 1476-88.

197.   Brühl, A., et al., *Neuroimaging in social anxiety disorder—a meta-analytic review resulting in a new neurofunctional model.* Neurosci Biobehav Rev., 2014. **47**: p. 260-80.

198.   Gómez-Pinilla, F., et al., *Voluntary exercise induces a BDNF-mediated mechanism that promotes neuroplasticity.* J Neurophysiol., 2002. **88**(5): p. 2187-95.

199.   Wu, A., Z. Ying, and F. Gomez-Pinilla, *Dietary omega-3 fatty acids normalize BDNF levels, reduce oxidative damage, and counteract learning disability after traumatic brain injury in rats.* J. Neurotrauma, 2004. **21**(10): p. 1457-67.

200.   Feinstein, J.S., et al., *The human amygdala and the induction and experience of fear.* Curr. Biol., 2011. **21**: p. 34-38.

201.   Rodrigues, S., J. LeDoux, and R. Sapolsky, *The influence of stress hormones on fear circuitry.* Annu Rev Neurosci., 2009. **32**: p. 289-313.

202.   Somerville, L.H., et al., *Interactions between transient and sustained neural signals support the generation and regulation of anxious emotion.* Cereb Cortex., 2013. **23**(1): p. 49-60.

203.   Sink, K.S., M. Davis, and D.L. Walker, *CGRP antagonist infused into the bed nucleus of the stria terminalis impairs the acquisition and expression of context but not discretely cued fear.* Learn Mem., 2013. **20**(12): p. 730-9.

204.   Arruda-Carvalho, M. and R. Clem, *Prefrontal-amygdala fear networks come into focus.* Front Syst Neurosci., 2015. **9**: p. 145.

205.   Likhtik, E. and R. Paz, *Amygdala-prefrontal interactions in (mal)adaptive learning.* Trends Neurosci., 2015.

206. Wood, J. and J. Grafman, *Human prefrontal cortex: processing and representational perspectives.* Nat Rev Neurosci, 2003 **4**(2): p. 139-47.

207. Southwick, S., M. Vythilingam, and D. Charney, *The psychobiology of depression and resilience to stress: implications for prevention and treatment.* Annu Rev Clin Psychol, 2005. **1**: p. 255-91.

208. Feder, A., E.J. Nestler, and D.S. Charney, *Psychobiology and molecular genetics of resilience* Nature Reviews Neuroscience, 2009. **10**: p. 446-457.

209. Chrousos, G., *Stress and disorders of the stress system.* Nat Rev Endocrinol., 2009. **5**(7): p. 374-81.

210. Bremner, D.J., *Gender Differences In Cognitive And Neural Correlates Of Remembrance Of Emotional Words.* Psychopharmacology Bulletin, 2001. **35**: p. 55-87.

211. Gorka, S., et al., *Cannabinoid Modulation of Amygdala Subregion Functional Connectivity to Social Signals of Threat.* Int J Neuropsychopharmacol., 2014.

212. Phan, K., et al., *Cannabinoid modulation of amygdala reactivity to social signals of threat in humans.* J Neurosci, 2008. **28**(10): p. 2313-9.

213. O'Connell, B., D. Gloss, and O. Devinsky, *Cannabinoids in treatment-resistant epilepsy: A review.* Epilepsy Behav, 2017. **70**: p. 341-348.

214. de Mello, S.A., et al., *Antidepressant-like and anxiolytic-like effects of cannabidiol: a chemical compound of Cannabis sativa.* CNS Neurol Disord Drug Targets, 2014. **13**(6): p. 953-60.

215. Zuardi, A.W., et al., *Action of cannabidiol on the anxiety and other effects produced by delta 9-THC in normal subjects.* Psychopharmacology (Berl), 1982. **76**: p. 245–250.

216. Zuardi, A., et al., *Effects of ipsapirone and cannabidiol on human experimental anxiety.* Psychopharmacol., 1993. **7**: p. 82–88.

217. Sotres-Bayon, F. and G.J. Quirk, *Prefrontal control of fear: more than just extinction* Current Opinion in Neurobiology, 2010. **20**: p. 231–235.

218. Anonymous, *Hospital Statistics.* 1999, Chicago: American Hospital Association.

219. Brennan, T.A., et al., *Incidence of adverse events and negligence in hospitalized patients: Results of the Harvard Medical Practice Study I.* N. Engl. J. Med., 1991. **324**: p. 370–376.

220. Starfield, B., *Is US Health really the best in the world?* JAMA, 2000. **284**: p. 483-5.

221. Kohn, L.T., J.M. Corrigan, and M.S. Donaldson, *To Err is human: Building a safer Health System.* 1999, Washington DC: Institute of Medicine.

222. Weingart, S.N., et al., *Epidemiology of medical error.* BMJ, 2000. **320**(7237): p. 774-7.

223. Thomas, E.J., et al., *Costs of Medical Injuries in Utah and Colorado.* Inquiry, 1999. **36**: p. 255-64.

224. Woolf, S.H., *Patient safety is not enough: targeting quality improvements to optimize the health of the population.* Ann Intern Med, 2004. **140**: p. 33-6.

225. Woolf, S.H., *The need for perspective in evidence-based medicine.* JAMA, 1999. **282**: p. 2358-65.

226. Heubi, J.E., M.B. Barbacci, and H.J. Zimmerman, *Therapeutic misadventures with acetaminophen: hepatoxicity after multiple doses in children.* J Pediatr., 1998. **132**(1): p. 22-7.

227. Lee, W., *Acetaminophen (APAP) hepatotoxicity—Isn't it time for APAP to go away?* Journal of Hepatology, 2017. **67**: p. 1324–1331.

228. Larson, A.M., et al., *Acetaminophen-induced acute liver failure: results of a United States multicenter, prospective study.* Hepatology., 2005. **42**(6): p. 1364-72.

229. Lee, W.M., *Acetaminophen and the U.S. Acute Liver Failure Study Group: lowering the risks of hepatic failure.* Hepatology, 2004. **40**(1): p. 6-9.

230. Møller, A.R., *Intraoperative Neurophysiologic Monitoring, 2nd Edition.* 2006, Totowa, New Jersey: Humana Press Inc.

231. Diekema, D., *Improving Childhood Vaccination Rates.* N Engl J Med, 2012. **366**: p. 391-393.

232. Carek, P., S. Laibstain, and S. Carek, *Exercise for the treatment of depression and anxiety.* Int J Psychiatry Med., 2011. **41**(1): p. 15-28.

233. Jain, A., K. Aggarwal, and P. Zhang, *Omega-3 fatty acids and cardiovascular disease.* Eur Rev Med Pharmacol Sci., 2015. **19**(3): p. 441-5.

234. Grant, W., *Insufficient sunlight may kill 45,000 Americans each year from internal cancer.* J Cosmet Dermatol. , 2004. **3**(3): p. 176-8.

235. Grant, W., R. Strange, and C. Garland, *Sunshine is good medicine. The health benefits of ultraviolet-B induced vitamin D production.* J Cosmet Dermatol., 2003. **2**(2): p. 86-98.

236. Aloia, J., et al., *Vitamin D intake to attain a desired serum 25-hydroxyvitamin D concentration.* Am J Clin Nutr, 2008. **87**(6): p. 1952-8.

237. Ruggiero, C., et al., *Omega-3 polyunsaturated fatty acids and immune-mediated diseases: inflammatory bowel disease and rheumatoid arthritis.* Curr Pharm Des, 2009. **15**(36): p. 4135-48.

238. Skulas-Ray, A.C., et al., *Omega-3 fatty acid concentrates in the treatment of moderate hypertriglyceridemia.* Expert Opin Pharmacother, 2008. **9**: p. 1237-48.

239. Maroon, J., C. and J.W. Bost, *Omega-3 fatty acids (fish oil) as an anti-inflammatory: an alternative to nonsteroidal anti-inflammatory drugs for discogenic pain.* Surg Neurol., 2006. **65**(4): p. 326-31.

240. Babcock, T.A., T. Dekoj, and N.J. Espat, *Experimental studies defining omega-3 fatty acid antiinflammatory mechanisms and abrogation of tumor-related syndromes.* Nutr Clin Pract, 2005. **20**: p. 62-74.

241. Blondeau, N.L., RH.;Bourourou, M.; et al, *Alpha-Linolenic Acid: An Omega-3 Fatty Acid with Neuroprotective Properties—Ready for Use in the Stroke Clinic?* Biomed Res Int, 2015. **2015**: p. ID 519830, 8 pages.

242. Osher, Y. and R.H. Belmaker, *Omega-3 fatty acids in depression: a review of three studies.* CNS Neurosci Ther., 2009. **15**(2): p. 128-33.

243. Mazda, J. and e. al., *Association between pre-diagnostic circulating vitamin D concentration and risk of colorectal cancer in European populations:a nested case-control study.* BMJ Clin Evid, 2010.

244. Manson, J., et al., *Vitamin D Supplements and Prevention of Cancer and Cardiovascular Disease.* N Engl J Med, 2018.

245. Manson, J., et al., *Marine n–3 Fatty Acids and Prevention of Cardiovascular Disease and Cancer.* New England Journal of Medicine, 2018.

246. Weedon-Fekjær, H., et al., *Breast cancer tumor growth estimated through mammography screening data.* Breast Cancer Res., 2008. **10**(3).

247. Araújo, J., et al., *Folates and aging: Role in mild cognitive impairment, dementia and depression.* Ageing Res Rev., 2015. **9**(19).

248. Das, U.N., *Folic acid and polyunsaturated fatty acids improve cognitive function and prevent depression, dementia, and Alzheimer's disease—But how and why?* Prostaglandins Leukot Essent Fatty Acids., 2008. **78**: p. 11-19.

249. Kondo, A., Morota, N, Date, H, Yoshifuji, K, Morishima, T, Miyazato, M, Shirane, R, Sakai, H, Pooh, KH, Watanabe, T, *Awareness of folic acid use increases its consumption, and reduces the risk of spina bifida.* Br J Nutr., 2015. **114**(1): p. 84-90.

250. Kelly, D., T. O'Dowd, and U. Reulbach, *Use of folic acid supplements and risk of cleft lip and palate in infants: a population-based cohort study.* Br J Gen Pract., 2012. **62**(600): p.:e466-72.

251. Martinez de Villarreal, L.E., et al., *Weekly Administration of Folic Acid and Epidemiology of Neural Tube Defects.* Matern Child Health J, 2006. **10**(5): p. 397-401.

252. Surén, P., et al., *Association Between Maternal Use of Folic Acid Supplements and Risk of Autism Spectrum Disorders in Children.* JAMA, 2013. **309**(6): p. 570-577.

253. Chitayat, D., et al., *Folic acid supplementation for pregnant women and those planning pregnancy - 2015 update.* J Clin Pharmacol., 2015. **13**.

254. Arth, A., et al., *Supplement use and other characteristics among pregnant women with a previous pregnancy affected by a neural tube defect - United States, 1997-2009.* MMWR Morb Mortal Wkly Rep., 2015. **64**(1): p. 6-9.

255. Long, D.M., *Chronic back pain,* in *Handbook of Pain,* P.D. Wall and R. Melzack, Editors. 1999, Churchill Livingstone: Edinburgh. p. 539-538.

256. Kahneman, D.T., A. (Eds.), *Choices, Values and Frames.* 2000, New York: Cambridge University Pres.

257. Thaler, R.H., *Quasi-Rational Economics.* 1991, New York: Russell Sage Foundation.

258. Charabi, S., et al., *Acoustic neuroma/vestibular schwannoma growth: past, present and future.* Acta Otolaryngol. (Stockh), 1998. **118**: p. 327-32.

259. Biller-Andorno, N. and P. Jüni, *Abolishing Mammography Screening Programs? A View from the Swiss Medical Board.* NEJM, 2014. **370**(21): p. 1965-7.

260. Wilkinson, J., *Effect of mammography on breast cancer mortality.* Am Fam Physician, 2011. **84**(11): p. 1225-7.

261. Welch, H. and H. Passow, *Quantifying the benefits and harms of screening mammography.* JAMA Intern Med, 2014. **174**: p. 448-454.

262. Baines, C., *Are there downsides to mammography screening?* Breast J., 2005. **Suppl 1**: p. S7-10.

263. He, F.J. and G.A. MacGregor, *Cost of poor blood pressure control in the UK: 62,000 unnecessary deaths per year.* J Hum Hypertens, 2003. **17**(7): p. 455-7.

264. Law, M., N. Wald, and J. Morris, *Lowering blood pressure to prevent myocardial infarction and stroke: a new preventive strategy.* Health Technol Assess., 2003. **7**(3): p. 1-94.

265. Kokubo, Y. and C. Matsumoto, *Hypertension Is a Risk Factor for Several Types of Heart Disease: Review of Prospective Studies.* Adv Exp Med Biol, 2017. **956**: p. 419-426.

266. Pistoia, F., et al., *Hypertension and Stroke: Epidemiological Aspects and Clinical Evaluation.* High Blood Press Cardiovasc Prev., 2016. **23**(1): p.:9-18.

267. Jetten, M., et al., *'Omics analysis of low dose acetaminophen intake demonstrates novel response pathways in humans.* Toxicol Appl Pharmacol, 2012. **259**(3): p.:320-8.

268.   Aminoshariae, A. and A. Khan, *Acetaminophen: old drug, new issues.* J Endod. , 2015. **41**(5).

269.   Fontana, R., *Acute Liver Failure including Acetaminophen Overdose.* Med Clin North Am., 2008. **92**(4): p. 761–794.

270.   Schultz, S., H. Klonoff-Cohen, and D. Wingard, *Acetaminophen (paracetamol) use, measles-mumps-rubella vaccination, and autistic disorder. The results of a parent survey.* Autism 2008. **12**: p. 293–307.

271.   Stergiakouli, E., A. Thapar, and G. DaveySmith, *Association of acetaminophen use during pregnancy with behavioral problems in childhood: evidence against confounding.* JAMA Pediatrics, 2016. **170**: p. 964–970.

272.   Gerth, J. and T. Miller, *Use only as directed.* Pro-Pblica, 2013.

273.   Beasley, R., et al., *Acetaminophen Use and Risk of Asthma, Rhinoconjunctivitis, and Eczema in Adolescents.* Am J Respir Crit Care Med, 2011. **183**: p. 171–178.

274.   Thompson, J., et al., *Associations between acetaminophen use during pregnancy and ADHD symptoms measured at ages 7 and 11 years.* Plos One, 2014. **9**: p. e108210.

275.   Walton, R. and W. Monte, *Dietary methanol and autism. Med Hypotheses.* Med Hypotheses, 2015. **85**: p. 441–446.

276.   Liew, Z., et al., *Maternal use of acetaminophen during pregnancy and risk of autism spectrum disorders in childhood: A Danish national birth cohort study.* Autism Res, 2016. **9**(9): p. 951-8.

277.   Kjellberg, F. and M.R. Tramer, *Pharmacological control of opioid-induced pruritus : a quantitative systematic review of randomized trials.* Europ. J. .Anaesthesiol., 2001. **18**(6): p. 346-57.

278.   Kuraishi, Y., T. Yamaguchi, and T. Miyamoto, *Itch -scratch responses induced by opioids through central mu opioid receptors in mice.* 2000. **7**(3): p. 248-52.

279.   Tohda, C., T. Yamaguchi, and Y. Kuraishi, *Intracisternal injection of opioids induces itch -associated response through mu-opioid receptors in mice.* Jap. J.Pharmacol., 1997.

280.   Hunt, S.P. and C.E. Urch, *Pain, Opiates and Addiction*, in *Wall and Melzack's Textbook of Pain*, S.B. McMahon and M. Koltzenburg, Editors. 2006, Elsevier: Amsterdam. p. 349-359.

281.   Porter, J. and H. Jick, *Addiction rare in patients treated with narcotics.* N Engl J Med, 1980. **302**(2): p. 123.

282.   Woods, J., *Abuse liability and the regulatory control of therapeutic drugs: untested assumptions.* Drug and Aicohol Dependence, 1990. **25**: p. 229- 233.

283.   Massey, E.W. and J.M. Massey, *Forearm neuropathy and pruritus.* Southern Medical J., 1986. **79**(10): p. 1259-1260.

284.   Zakrzewska-Pniewska, B. and M. Jedras, *Is pruritus in chronic uremic patients related to peripheral somatic and autonomic neuropathy? Study by R-R interval variation test (RRIV) and by sympathetic skin response (SSR).* Neurophysiol Clinique, 2001. **311**(3): p. 181-93.

285.   Franklina, K., *Analgesia and Abuse Potential: An Accidental Association or a Common Substrate?* Pharmacology Biochemistry and Behavior, 1998. **59**(4): p. 993-1002.

286.   Marengo, A., C. Rosso, and E. Bugianesi, *Liver Cancer: Connections with Obesity, Fatty Liver, and Cirrhosis.* Annu Rev Med., 2016. **67**: p. 103-17.

287.   Wu, D. and A. Cederbaum, *Oxidative stress and alcoholic liver disease.* Semin Liver Dis., 2009. **29**(2): p. 141-54.

288.   Pelucchi, C., et al., *Alcohol consumption and cancer risk.* Nutr Cancer., 2011. **63**(7): p. 983-90.

289.   Bagnardi, V., et al., *Alcohol consumption and site-specific cancer risk: a comprehensive dose-response meta-analysis.* Br J Cancer, 2015. **112**(3): p. 580-93.

Subject Index

V

Made in the USA
Las Vegas, NV
18 May 2022